A Lifetime of

Beekeeping

Mistakes

———

GEOFF CRITCHLEY

5m Books

First published 2023

Published by
5m Books Ltd
Lings, Great Easton
Essex CM6 2HH, UK
Tel: +44 (0)330 1333 580
www.5mbooks.com

A Catalogue record for this book is available from the British Library

ISBN 9781789182026
eISBN 9781789182361
DOI 10.52517/9781789182361

Book design and layout by Alex Lazarou

Printed by CPI Anthony Rowe Ltd, UK

Cover photo: hedera.baltica (Flickr)

Pictures and diagrams by Geoff Critchley except:
Fig 1.1, 1.2, 2.1, 2.14, 5.1, 5.2, 6.1, 7.2, 8.1, 9.1, 11.1 and 13.1 by
Asher Blythen, Fig 8.2 Natalia Melnychuk / Shutterstock,
Fig 8.5 J-R / Shutterstock, Fig 8.6 Tasnenad / Shutterstock,
Fig 8.7 Fabio Merelli / Shutterstock, Fig A4.1 Ihor Hvozdetskyl /
Shutterstock, Fig A4.2 kosolvskyy / Shutterstock

For
My granddaughters –
Amelia, Bronwen, Emily and Neve

Contents

About the author

Geoff, born 1949, is originally from Somerset, where as a boy he was introduced to bees at school. After moving to North Wales in 1984 he started keeping bees. He is a Master Beekeeper and has been an examiner for the British Beekeeping Association. For many years he taught beginners for Conwy Beekeeping Association and South Clwyd Beekeeping Association, and has run training courses on queen rearing. Geoff at one time had as many as 60 hives and worked for 3 years as a seasonal bee inspector for the National Bee Unit. He now has just two hives but continues to mentor new beekeepers.

Glossary

14 x 12 hive Similar to the National hive (see below) but with 11 deep brood frames, each measuring approximately 14" × 12" (355 mm × 304 mm).

Asian hornet, *Vespa velutina* A non-native hornet from southeast Asia, now a major problem in France, although not yet established in mainland UK.

brood and a half This is a system used by many beekeepers in UK. It is a double brood box hive where one box is a standard national depth 8.5" (215 mm) together with a second brood box the size of a standard super – approximately 6" (150 mm) deep.

cast A small swarm with a virgin queen, normally produced 1 week after the first swarm from a colony. Sometimes referred to as a secondary swarm.

Defra The UK Department for Environment, Fishing and Rural Affairs. The National Bee Unit, who are responsible for checking and dealing with bee diseases were part of this organisation. The Bee Unit is now part of Fera, Food and Environment Research Agency.

mini nuc A mini nuc is a very small hive only large enough to contain only a cup full of bees together with a virgin queen. Sometimes referred to as a mating nuc. Normally only used in large-scale queen production.

National hive The most commonly used hive in the UK with 11 brood frames, each measuring approximately 14" × 8.5" (355 mm × 215 mm).

nuc A nuc, or nucleus, hive is a small size hive normally containing just 5 or 6 frames.

propolis Manufactured by the bees from saps and resins from plants and trees. It is used to glue the hive together and to fill up small gaps. It has natural antifungal, antibacterial and antibiotic properties. Some bees use it to reduce the size of the entrance, which is where its name is derived from, being Greek meaning defender of the city.

queen cage A small wire or plastic cage used to contain the queen when introducing a queen into a hive of bees.

small hive beetle A beetle from sub-Saharan Africa. Now a major problem for beekeepers in the USA.

supersedure A method used by bees to raise a new queen to replace an older queen, without swarming taking place.

travelling cage Similar to the queen cage, but used to transport a queen together with about 6 worker bees and a small quantity of sugar fondant.

Varroa A parasitic mite, *Varroa destructor*, originally a mite on the Asian honeybee (*Apis cerana*) known as *Varroa jacobsoni*, which then crossed the European honeybee (*Apis mellifera*).

WBC hive The classic cottage garden hive. Uses the same frames as the National hive, but has only 10 brood frames.

Introduction

This book grew out of a talk that I have given to local beekeeping associations, which started as '25 years of beekeeping mistakes', grew to 30 years and then to 35 years. At the end of each talk I would ask the audience if anyone had not made at least one of the mistakes I had mentioned. No one ever admitted to never having made a mistake!

I have kept bees for almost four decades, but it is 60 years since my first introduction to honeybees. During this time, I've owned up to 60 hives, processed and sold tonnes of honey. I've made lots of mistakes, most of which I'll be owning up to. I've worked as a seasonal bee inspector, so have had the opportunity that is denied most beekeepers, that of being able to look at other people's bees and their equipment. From that experience I learnt a lot and saw things that didn't always comply with my concept of best practice.

This book is not meant as a beekeeping textbook. Rather it contains examples of mistakes that shouldn't, in my opinion, really be repeated. It also has guidance on my idea of the correct way of doing things. I know it will not accord with some other beekeepers' ideas because there is probably no absolute right or wrong way to keep bees. If it works for you and you are happy with it then why change, unless it has an impact on your neighbours or other beekeepers.

Our beekeeping is influenced by where we live, which in my case is about 900 ft (275m) above sea level on the east side of a mountain in North Wales. That should disprove the myth that you can't keep bees above 600 ft (180 m). My way of beekeeping will not therefore be appropriate for a beekeeper in an arable region in the south of England, where the forage and climate are somewhat different, but the same principles should apply.

I've become interested in our native honeybee *Apis mellifera mellifera*, and make no secret of my opposition to importation of bees. My comments and observations may well differ from that of other beekeepers, but that is to be expected.

Like many other beekeepers, I have experimented and sometimes have got it wrong. Often, I thought what I had been doing was correct, only to find there is another, often better, way. We have to be open to suggestions, to try new ideas and be prepared to make mistakes. I remember the advice that my father gave when I was a teenager – 'He who has never made a mistake has never made anything.' I discovered much later this is actually a misquote from Albert Einstein 'Anyone who has never made a mistake has never tried anything new.' There are different categories of mistake, some are just accidents, some are things about which with hindsight we should have known better, others are just caused by a lack of knowledge.

What I have learnt is that when you keep bees you start watching the changing seasons and looking at the environment in greater depth. Our honeybees, along with so much of nature, are under pressure, most of which is caused by us humans. Humankind's biggest mistake was to take its eye off the ball with regard to the climate. The pressure to produce food has irreversibly changed our countryside. The genie is out of the bottle and it will have impacts on our beekeeping and us forever. We can look at the mistakes of the past, and hopefully learn from them. Our beekeeping changed when we had to deal with Varroa, and it will change again when we have to deal with the small hive beetle and the Asian hornet. The next generations of beekeepers will have to be flexible to also deal with changes to the countryside and climate.

1

Starting out

My introduction to beekeeping came during my second year at grammar school, in Somerset. Our science teacher was Rex Sawyer, who was a leading beekeeper in the county at that time. When he retired from teaching, he went on to research pollen identification at Cardiff University and wrote highly acclaimed books on pollen and honey identification. I feel privileged to have known him. He kept some hives on the flat roof of the science laboratory. Health and safety would not allow what happened in those days, but back in 1962 it was perfectly acceptable for a class of 13-year-old boys, possibly some in short trousers wearing simple veils, to

1.1
Don't worry boys they won't sting!

climb a ladder and gather round a hive of bees, while our teacher opened up the hive. How things have changed! The honey was 2s 6d (half crown, or 2 shillings and 6 pence, roughly 12.5 pence) a jar, although I think there was a limit of four jars per pupil.

I really didn't give bees much more thought until 1984 when we moved to North Wales to a house with a large garden on the edge of a village. We wanted to use part of the garden for growing vegetables and started keeping a few chickens in an attempt to engage with the environment. What could be better than our own home-grown vegetables, free range organic eggs and home-produced honey? Late in the summer of 1985 I spotted an advert in the local newspaper for bees for sale, and that was the start of my beekeeping. Many years later, when I was running beekeeping courses, I asked students why they wanted to keep bees. Very few wanted to do it to make money from selling honey. Most wanted to embrace the natural world, or to save the bees that they had heard were in danger of extinction. If I ever re-asked the question years later the replies were different. Then the common response was that they hadn't realised how interesting the hobby was going to be. Some of course never achieved their expectation and gave up. I suppose the mistake for them was even to have thought of keeping bees. I suspect that sometimes my wife thinks that my first mistake was to want bees!

When I obtained my first hive, apart from reading a book on beekeeping I knew nothing. The bees were bought and moved to an apiary site in a large garden on the outskirts of the village. I wasn't confident about having bees in my own garden and was fortunate to be offered a site by an old lady who was happy to see her overgrown garden being put to better use. I knew nothing about selection of protective clothing and didn't realise that dark blue overalls and trouser legs tucked into dark woolly socks was not the best idea! I subsequently learnt that bees don't like dark colours and get tangled in woollen clothes. I also learnt that stings hurt, and I suffered swollen ankles for a few days.

I was comforted by the unproven knowledge that bee stings were good as a treatment for painful joints, and perhaps the short-term pain would give me lifelong relief from stiff and painful knees. Now aged 72

I wonder how much worse my joints would be if it weren't for regular bee sting therapy!

This first colony of bees made it through the winter. It was much easier in those pre-Varroa days when, if you lost 10% of your colonies you had done badly. These days if you only lose 10% over winter you can think yourself lucky.

I joined the local beekeeping association, but really beekeeping was still a self-taught hobby. There were apiary meetings on Saturday afternoons, but when working all week I found there were other important things to do at weekends. I learnt nothing from those meetings. It seemed to me to be just opening up a few generally uncared for hives and trying to find the queen at the same time as trying to keep the smoker alight. Most inspections failed to find the queen. There were no record cards. I do remember some of the bees being a bit grumpy, and they certainly followed us out of the apiary when we had finished. The apparent highlight of the meeting was the tea and cakes afterwards. In those days it felt as if you could keep a colony of bees and take some honey off at the end of summer, feed them some sugar for the winter and they went on from year to year however good or bad a beekeeper you were.

Things are very different now, and in recent years there has been a resurgence of interest in beekeeping, but it has got more difficult. It is now the latest thing for celebrities to try, and then tell everyone else on social media how marvellous a hobby it is. Time will tell if they all keep it up.

During my first full year of beekeeping I was once stung on the back of my right hand, near the base of my ring finger. The hand swelled up, and looked like a blown up rubber glove. It required a visit to the local minor injuries clinic. The nurse had to cut off my ring and was very sympathetic about the sting. She asked if I was sure it wasn't a wasp that had stung me. I replied that I was sure it was a bee as it was one of mine! At that point, for some reason, all sympathy disappeared.

I live on the boundary of two associations and after a few years, I changed my allegiance, and joined the other association as they were a larger group, had good speakers during the winter months and were generally more active.

Many years later, this association which had been providing introductory courses for beginners for a few years asked me to take over running the course, and that led on to me running summer, winter and weekend courses for two North Wales associations as well as weekend residential courses at home, with students in some cases coming from over 250 miles (402 km) away. I often wonder how many of those would-be beekeepers still keep bees or even if they ever started up, following my introduction to getting up close to these stinging insects.

There are now beekeeping courses, run by local associations, all over the country and I always advise potential beekeepers to contact their local association and enrol on a course, to go to apiary meetings and be sure they know what it is they are taking on before actually getting their own bees. Find a bee buddy to help you. Many associations have schemes to help and support new beekeepers. You don't have to struggle on your own, blundering about assuming you have got it right when all along you are making work for yourself – as was the case when I started.

You don't have to buy new equipment, but exercise caution and seek advice before buying on the internet. I've seen some good second-hand equipment for sale and some absolute rubbish. Looking back, I suppose the equipment that came with my first bees was not the best, but I was green and wanted to keep bees! When purchasing second-hand hives we have to avoid diseases, but I'll cover that in Chapter 8. When you buy your bees, my recommendation is to buy local and my reason for this is because you never know what you're getting from some suppliers, again I'll explain this in more detail later in the book.

Once, when giving a talk to a local Women's Institute meeting, I showed a slide of me with a hive tool and a smoker and foolishly said that those two tools are really the only equipment needed. My wife had accompanied me that evening and was sitting at the back of the room. She was rather quiet on the journey home and was obviously pondering something. She said, 'I've been thinking – if all you need is a hive tool and a smoker, why do you have two sheds full?'

Later when giving a list of beekeeping equipment to a class of beginners I jokingly included two sheds as essential. Of course, no one took this as anything other than a joke, but a few years later, one of the learners,

1.2
'Why does Geoff have two sheds?'

now an established beekeeper said to me, 'You know you said we would need two sheds. None of us believed you, but guess what?'

Lastly when starting out we have to decide where we are going to keep our bees. Unless you have a very large plot of land, or no near neighbours, you may be better looking for an apiary site close by, as I did. Some farmers are only too pleased to have bees on their farms, particularly if they are part of an environmental management scheme, while others may not be interested. Once you have mastered the art of keeping bees then you can consider having some in your garden.

Even when I had more than 60 hives of bees I usually had just a few hives at home. They were located in an area of the garden fenced off behind a 6 ft (2 m) high fence. They never caused any problem to us or to our neighbours. Everything changed one day when there was a knock on the door, it was our new neighbour who had moved in only a few weeks earlier. They were worried about their children being stung, as the fence on their side of the apiary was only 3 ft (1 m) high, but was backed by a substantial laurel hedge. I spent the next day replacing the short fence with one 6 ft (2 m) tall. It was a burning hot day and dressed in a bee suit so that I could work close to the hives I proceeded to replace the fence, digging out and re-concreting the posts. It took all day to replace three fence panels and I was dripping with perspiration. My wife thought I was going to have a coronary; I didn't of course. The children didn't get stung, and over the years the neighbours have got used to the fact that these deadly insects aren't so bad. You have to modify your beekeeping though, particularly if the children's trampoline is just over the hedge! The bees have to fly up over the fences and then they are above head height and cause no problem either in our garden or the neighbours'. You just have to choose the right time to look in the hives and hope the children don't come home when you are halfway through an inspection.

If I hadn't had a few years' experience in keeping bees I would have fallen out with the neighbours and/or had to move the bees away.

If you are setting up an apiary away from home, make sure you can get there in all weathers, all year round. That is true of temporary apiary sites as well. A few years ago, I took some bees to Herefordshire. I shared the site with a local commercial beekeeper on a field of borage. Once the crop had finished, I transported the bees back (more of that later). The other beekeeper left the hives there for a few weeks and the weather turned wet. The ground became soft and he couldn't get a vehicle near the hives until the spring of the following year.

One thing about apiary selection that is often referred to in books is not to put hives underneath trees. The reason given is that the rain dripping from the tree will disturb the bees. That reason didn't make a lot of sense to me, and often an out apiary site has to be a compromise. Landowners don't want an apiary set up in an open area of a field, but are normally quite happy for it to be in a corner, where it is difficult to mow or plough. Often there are overhanging trees or an overgrown hedge. This can be an advantage as it provides partial shade from hot summer sun, and shelter from wind and rain. There are disadvantages too. I have experienced two incidences that no book that I had read ever mentioned. One winter a small hawthorn tree in the hedgerow behind my hives blew down in a storm and fell on top of two hives. Using a chainsaw, while wearing a bee suit, is interesting. Which PPE does one choose – hard hat, visor and ear defenders, thick leather gloves, chainsaw proof trousers and boots, or bee suit? On this occasion both hives had been strapped together so no harm was done. The other occasion was caused by a summer storm. A branch of an ash tree had broken, and swung down knocking two hives over, smashing the hive stand. I had to cut up the branch with a chainsaw just to get to the hives and then set up

"Using a chainsaw, while wearing a bee suit, is interesting. Which PPE does one choose – hard hat, visor and ear defenders, thick leather gloves, chainsaw proof trousers and boots, or bee suit?"

a new hive stand and put the hives back together. One colony survived the trauma but the other never recovered, and the hive was only partially reusable. Unfortunately, on both these occasions, having cut up the wood I never even got the logs to burn on the fire.

I once had some bees on a local fruit and vegetable farm. The bees were located in an uncultivated corner, which happened to have overhead power lines running across it. My relatively docile bees became quite bad tempered, which at the time I put down to the overhead electricity lines. I've since seen anecdotal reports that bees are affected by the magnetic fields set up by power cables. I moved the hives away after one season and they seemed better tempered on their new site on an organic farm. More recently, I've taken over some hives on a site where there are power lines almost directly above the hives and they are quiet, good-tempered bees. Perhaps the overhead electricity cables have no affect after all.

There are lessons here. Before setting up your first apiary, seek advice of other beekeepers. Speak to your neighbours!

During 1997 I decided to study for the British Beekeeping Association (BBKA) exams, and over the next few years passed the basic assessment and then the theory and practical exams to become a master beekeeper. It was only by studying bees in depth that I learnt to understand what was going on in the hive, and why we need to do what we do and when to do it. It allowed me to look again at what I was doing and improve my beekeeping.

At one time during my semi-commercial phase of beekeeping, I became the seasonal bee inspector for northeast Wales, working for Defra. In a typical season I would visit about 150 beekeepers and open about 600 hives. In that time, I saw examples of very good beekeeping, but sadly that is not true of all beekeepers. During those 3 years that I spent in that job I learnt many things that improved my own beekeeping and identified many things to avoid doing. I certainly became very confident about opening up a hive of bees that I had not previously seen, and had no idea of its temperament. Many mistakes described later in the book relate to that time, rather than my own errors. Unfortunately, one farmer, on hearing that I was working for Defra as an inspector, asked me to take my bees off his land. For some reason he didn't want a ministry official coming onto his farm! The following year I inspected bees belonging to another beekeeper on this same farm.

The last piece of advice I give to potential beekeepers is to remember most people have little knowledge of bees; they only recognise bumble bees and think honeybees are wasps! This results in the new beekeeper becoming the local expert on all flying insects. Expect to be asked about removing wasp nests, bumble bees and solitary bees.

R.O.B. Manley in his 1948 book *Beekeeping in Britain* wrote 'When I began ... ... I constantly ran up against difficulties that no book that I could find did much to explain. The authors often contradicted one another, and which of them was a poor novice to believe.'

I don't think anything has changed in the intervening 70 years!

Perhaps I was lucky in those early years of my beekeeping, but I read all the beekeeping books there were in the local library (actually both books – it wasn't the most popular topic in the non-fiction section), and I had my reference book – *A Guide to Bees and Honey* by Ted Hooper.

Years later I'm still surprised by the questions that I've been asked by novice beekeepers. Some years ago, I was advertising native queens for sale on my website, and was telephoned one day by a man wanting a queen. My queens were in short supply and I really only wanted to sell them to someone who wanted native bees. I asked why he wanted a queen, expecting the normal reply that he wanted to change over to native bees as he wasn't happy with his angry mongrel bees, but I wasn't anticipating the answer I got. 'Well I've got a hive so I need a queen.' It took a few seconds for it to dawn on me that this man had no idea how to start keeping bees. The conversation went from bad to worse, of course, it transpired that he had already bought two queens from a commercial supplier and 'introduced' them together with the few attendant workers that came with them into his hive and he couldn't understand why they had died on both occasions!

I have been asked to supply a nuc (nucleus colony) to beginners on many occasions and have said they would be available late May or more likely sometime in June (things never get going very early in the year when you live halfway up a mountain in North Wales). Twice when I contacted the beginner to say their nuc is ready they have told me they couldn't wait and they had already bought one from a supplier somewhere in the south of England, who had some available earlier than mine. The following year they asked me again for a nuc. 'How are your other bees doing?', I asked. Answer, 'They died.' You might wonder if I then supplied them with replacement nuc. I did but they weren't top priority on my list.

When I supply a nuc to a new beekeeper I always give them detailed instructions on how to proceed. Normally they follow the instructions, but sometimes they must think that the details are not that important. I know that some of the nucs that beekeepers buy from elsewhere have only been made up a day or two before delivery/collection. Ideally the brood and bees in the nuc should be from the queen in the hive, or at the very least the queen should be actively laying in the hive. I have heard of people who have received a nuc with a queen still in a cage, and on one occasion that queen was dead. What chance is there for a beginner when that is the start that they get?

I remember once, as a seasonal bee inspector, visiting a beekeeper who had purchased a nuc 2 weeks previously. I was surprised how little activity there was outside the hive, but I was even more surprised by how little activity there was inside the hive. The bees were on three frames sitting in the middle of a brood box with four frames of foundation either side. The bees were starving! I often ponder whether some people deserve to be trusted to look after these wonderful creatures.

I know that a nuc that I supply meets the requirements of at least three frames of brood and two more frames with stores, with plenty of bees. The recipient is told that they need to be transferred to a full hive within 48 hours, and then fed so that they will draw out the combs of foundation. I often follow up the sale with a phone call a few days later to see how things are going. I recall once selling a nuc and just over a week later being told over the phone that the bee inspector was coming in a few days' time and he had offered to help transfer the nuc to a full hive. I knew the bee inspector very well and I asked him how things had gone. He told me that the bees had swarmed by the time he got there. Ultimately, they recovered, expanded and lived through the following winter, but wouldn't it have been better if they had been put into a full-size brood box straight away and given the space they needed.

Other new beekeepers have wondered why the nuc has not drawn out the foundation and expanded to occupy the whole of the brood box. They hadn't got that vital bit of beekeeping equipment – a feeder.

I know of at least one new beekeeper who never managed to get any honey crop in the first 2 years of keeping bees. The bees had never drawn out the foundation in the super, because they had never been through the queen excluder. Bees are very loath to go into a super of foundation through a queen excluder. Beginners only ever have frames with foundation and when putting on the first super it is essential to put it straight on top of the brood box with no excluder. Once the bees have started to draw out the comb then the queen excluder can be put on, checking of course that the queen is in the brood box not in the super. They then will continue to move freely through the excluder.

Most new beekeepers, like me when I first kept bees, want to have plants and trees in their gardens that will be beneficial and help the bees to produce more honey. I read all I could about which plants bees like and knew that honeybees are used as pollinators on fruit trees. When planning our garden we always tried to choose plants that attracted bees, and we planted apple, pear and plum trees. You would have thought that with four hives in the garden, and at least ten other hives that I knew of within about a mile radius, that the garden would be full of bees. Apart from a cotoneaster bush, which is alive with bees when in flower, the rest of the trees and shrubs attract very few honeybees – plenty of bumble bees, solitary bees, butterflies and other insects. The biggest disappointment is how few honeybees are seen on the fruit trees. When these are in flower the honeybees are gathering nectar and pollen from sycamore and horse chestnut trees, or from dandelion which is abundant in some of the local fields. The mistake I suppose is to expect the bees to forage where we think is best for them.

2

Hives and tools

The most fundamental item of equipment is rather obvious – the hive! A quick glance in the catalogue of any of the beekeeping equipment suppliers will show you a wide array of different hives and frames to go in them. For someone starting out this can be a minefield. It is made even worse by the internet, with offers of "starter sets".

2.1
'WBC is best' – 'No, National is best.'

I've been approached on a number of occasions over the years by newcomers asking for bees. Sometimes the enquiry is because they have been given a hive as a Christmas or birthday present. There are two possible problems in this scenario. The first one is that the hive is different to any hive that is used locally. Most beekeepers that I know use either the National hive or the WBC, which both take the same frames. It is easy in this case to supply a nuc of bees that will fit into a hive. Unfortunately, this simple situation is not always the case. I agreed to supply bees to one new beekeeper only to find that the hive he had been given for Christmas was a Langstroth. The second problem is that the new hive or frames have not been correctly assembled and the necessary gaps between parts are wrong. The bees subsequently glue all the bits together with propolis and it is almost impossible to take the hive apart for inspection.

There is a growing movement in the UK for 'natural beekeeping'. This normally means using a hive with no frames, such as the Warre hive or some other form of hive that can't easily be inspected. You may well ask why it is necessary to inspect. Two reasons come to mind, first, to check for disease and, second, to control swarming. Natural beekeeping appears to me to rely on swarming as a means of populating a hive and this in turn might explain why these natural beekeepers claim that they don't need to treat for Varroa. (Don't have to or can't?) Some don't even take off any honey. I've been asked on a number of occasions if I can supply a nuc for a Warre hive, but the only way to populate one of these hives is by introducing a swarm, as the hive has no frames.

The other difficulty with keeping bees in that type of hive is the problem with swarming, particularly if the hives are in an urban location. As responsible beekeepers, surely, we should try to minimise the disruption caused by swarms of bees.

Another type of hive unsuitable for keeping bees in the UK is the Top Bar Hive, or sometimes referred to as the Kenyan Top Bar Hive. This hive was designed for use by beekeepers in Africa, as it could be made using any locally available materials. It doesn't have frames, just top bars on which the bees build comb. In Africa, the bees behave very differently to our European bees. They regularly abscond in what might be described as a migratory swarm when forage runs out. Therefore, the hive doesn't need a large capacity. To the uninitiated, this type of hive might offer a cheaper alternative to the rather complicated and expensive National, but it just doesn't fit the requirements for the European honeybee.

When trying to manage Varroa in particular, we surely need a modern hive that allows us the flexibility to do so. Those hives that were developed for use in rural Africa, or designed in the late 19th century, made using the minimum of woodworking skill, to my mind, don't fit the bill.

The National hive is not ideal, but it is used by the majority of British hobby beekeepers. I'm not a fan of the WBC hive, although it is used by many beekeepers probably because it is pretty and is seen as the traditional idea of a beehive.

The concept of bee space has guided the design of modern beehives and was first recognised and promoted by the Philadelphia minister Lorenzo Langstroth in 1851, when he introduced what is now commonly known as the Langstroth hive. He discovered that bees build excess comb in a space larger than 9 mm (3/8"). Bees will fill any space less than 6 mm (1/4") with propolis. This hive is widely used around the world, but not popular with hobby beekeepers in the UK. The concept of bee space is, however, universally adopted in all modern moveable frame beehives. An interesting feature that varies between hives is what is referred to as either top or bottom bee space. Top bee space is where there is a space above the frames, whereas bottom bee space has the top of the frames flush with the top of the box, either brood box or super. Bottom bee space appears to be universal to English hives, that is, National, WBC and Commercial, all other hives are top bee space – Langstroth, Dadant and the Smith hive, which is a Scottish design.

It is possible to have top bee space in a National, either intentionally by design or by incorrect construction of the hive. If you mix boxes of top and bottom bee space you will either have too big a gap or no gap, either way the bees will have filled the gap with comb or glued everything together, making it difficult to disassemble the hive.

Having acquired hives from many different sources I have found that there are slight differences between manufacturers, and in the way hives are put together. I don't know how many of these older hives were home-made, but sometimes the concept of bee space was not strictly adhered to, resulting in real problems with parts being stuck together with propolis. As a result of struggling over a number of years, I think I have become fixated on the need to get bee space correct. A hive that has the correct bee space everywhere is a joy to work with. Badly designed or badly assembled hives and frames can make inspection a nightmare.

One aspect of hive design is the use of a frame runner. These runners are designed to minimise the contact area between the frame lug and the hive body. Normally these runners are made from galvanised or stainless steel, but they can be made from plastic at about half the price of steel. Why then doesn't everyone use plastic? I suppose the answer only comes from experience of using both.

With wooden hives, the standard method of cleaning and sterilisation is to scrape off the propolis and then scorch with a blowlamp. Obviously, you can't do that with plastic! The other problem is strength, particularly when used in the brood box. The runner is fixed in place with a few small nails and is slightly bendy, so over time with the weight of the frame it can sag, particularly when it becomes softened by propolis. As soon as this happens, the frames no longer hang straight and the bee space is compromised. Plastic runners tend not to provide sufficient bee space anyway, so propolis under the frame lug becomes a problem. My recommendation is to not follow my mistake in thinking plastic was better. It is an inferior product. Use galvanised steel, or if you are not worried about the cost buy stainless steel at twice the price. Beware if you are buying seconds quality hives as they may come with plastic runners, so check before you buy and if necessary buy separate metal runners.

"As a result of struggling over a number of years, I think I have become fixated on the need to get bee space correct."

2.2
The correct bee space on a National hive. The top of the frame is just below the top of the sides. There is a full bee space underneath the frame lug and between the side bar of the frame and the inside of the box. There is only a small gap between the end of the frame lug and the box to ensure the self-spacing frames abut correctly.

2.3
Is this a design or manufacturing error?
It is actually an eke to convert a National to 14 × 12, supplied with plastic runners, but when fitted with a metal runner the frame sits proud of the top of the box. It was purchased as a 'seconds' quality, so perhaps that was the reason it was a second! The queen excluder won't sit down correctly on the top of the box.

2.4
This National hive is supplied without a metal or plastic runner. The top of the side timber is machined to a chamfer. This might be cheaper, but the bees do build more propolis around the joint of the frame and the runner. The box has also been assembled wrongly, as the top of the frame sits well below the top of the box, and there is less than a bee space below the frame lug. In use, this box becomes a beekeeping nightmare once the propolis builds up.

2.5
This National box is another with the timber machined to provide the runner. It has a very large contact area between the frame and the runner, so is prone to building up propolis. The box has also been put together wrongly, as the frame sits too high.

Some years ago, I assisted a charity group that helped homeless people. We set up an apiary as part of a gardening scheme and had funding to buy some hives and bees. Workshops were held on putting the hives and frames together, where I emphasised the importance of correctly gluing and nailing everything together. Then I just let them get on with it. Towards the end of the first year, I noticed that the bee space had changed at the top of the hives. Closer examination showed that the walls of the brood boxes had slipped. They hadn't been nailed together as per instructions but instead the boxes were only partially nailed and glued together with 'No More Nails'. This glue is not suitable for external use and over time, the boxes had come unglued. We had to go through each hive in turn, transfer the frames to a new brood box and then take the boxes apart and fully re-nail them.

One item of the hive, which I found gave me trouble in my early years, was the queen excluder. This simple piece of kit is in essence a screen with gaps large enough to let workers pass, but small enough gaps that the queen can't squeeze through. These queen excluders were once slotted zinc sheets; these days they are made from slotted galvanised steel. The zinc queen excluders were quite flexible and the slots could become bent or broken which would allow the queen through. The other problem is that the queen excluder is placed directly on top of the frames (bottom bee space hives), and the bees then propolised the queen excluder to the frames. When the excluder was put back on after an inspection, it didn't quite fit, so became slightly bowed and was thus partially filling the bee space. When the hive was opened the next time, the problem was even worse than the first time. Peeling back the queen excluder caused some disturbance to the bees.

> "One item of the hive, which I found gave me trouble in my early years, was the queen excluder."

2.6–2.7
The other problem with these excluders can be seen from these pictures. There is actually very little clear opening left for the bees. It does not make much difference which way round the excluder goes either.

It was almost by accident that I discovered the wire queen excluder. This type is framed with timber, giving a proper bee space on the lower side (on a bottom bee space hive). This excluder was a revelation. No more propolisation of the queen excluder, less disturbance of the bees, plenty of slots for the bees to go through, it is more robust and less prone to breakage.

If this type of queen excluder is so much better, then why wasn't it being universally used? Probably because it costs about three times the price of the cheaper slotted excluder, and most beekeepers are averse to spending more than the absolute minimum. Perhaps like me they have never experienced anything other than the slotted queen excluder.

2.8
Wire queen excluder.

2.9
Bee space above the queen excluder to the first frame in the super.

Some years ago, I came across plastic excluders. These were similar in design to the slotted zinc excluders. They were made of a flexible plastic which the propolis softened in use and caused the sheet to bow even worse than the zinc (plastic was the cheapest excluder at the time but not the best). Plastic excluders that I have seen since are made from a different type of plastic, and I've no experience of how these perform in practice but I've no intention of trying them.

In those early beekeeping years, solid floors were the norm. These solid floors of the past have now been widely replaced by open mesh floors (OMFs), which have a removable slide used to monitor Varroa mite drop. All of my hives have these OMFs and the floor is left open, all year round. The slide is only ever used to monitor mite drop.

A word of warning here! Don't leave the slide in for prolonged periods as it will collect wax cappings from the hatching brood and become a breeding ground for wax moths!

Mostly the bees leave the floor alone, but some hives do propolise the mesh. One problem I did encounter when I first changed to OMFs was the width of the sides of the floor. These particular floors, which I purchase readymade, had side bars that were much wider than the hive sides. As the frames were bottom bee space this shouldn't have made any difference, but in one hive the frames sat a little low in the box, as it had been incorrectly put together (not by me incidentally). The bee space at the bottom of the frame was thus too small and the bees propolised the frames to the edge of the floor.

The other problem when changing over to these OMFs was that there was burr comb along the bottom of the frames, which had to be removed before the frames would sit clear of the sides of the floor.

I learnt by experience that I needed to modify my hive stands when using OMFs so that returning foragers weren't attracted under the floor rather than to the hive entrance.

On one occasion in early evening, I was going through one hive to find the queen. I couldn't find her and it had become quite overcast and the light wasn't good, so I closed the hive up and went back the following day to look again. I still failed to find her so left the search for another day. Back the next day and still had no luck, but noticed there appeared to be very few eggs. There were no signs of queen cells, so the queen had to be there. Next thing to look at in these situations is in the supers, to see if she had slipped up there. No sign of the queen laying eggs in the super, so, where was she? Looking down I noticed many bees hanging under the OMF, almost like a small swarm. At times like this, a spare floor is always useful, so I carefully transferred the hive onto a new floor so that I could pick up the old floor for a look. Sure enough, there was the queen with a cluster of workers underneath the open mesh floor. She must have dropped off a frame when I was looking for her, and just ended up under the floor. The colony still thought she was there, but she couldn't get up into the hive.

I've seen this on another occasion when a swarm went into an empty hive, but the queen and many of the workers were under the floor and proceeded to make comb there.

This taught me that I needed to modify my hive stands so that returning foragers weren't attracted under the floor and went correctly to the hive entrance, by ensuring that the entrance is about 150 mm (6˝) above the bottom of the hive stand frame.

I was once helping a new beekeeper who wanted to make his own hive, so I lent him some hive plans for a National hive, and plans for an OMF. I supplied him with a nuc, which he transferred to his new hive. About a month later he asked me to have a look at the bees for him, just to reassure him that he was doing things right. We had a look in the hive and all seemed well, but then I looked down

2.10
Hive stand showing depth of timber below the entrance.

through the brood box. The OMF had no mesh at all, just an open hole. Never assume that instructions are clear.

The next part of the hive, which needs some thought, is the entrance block. I know many beekeepers don't use entrance blocks and just have a wide-open entrance the full width of the hive. Some of the old beekeeping books have guidance on entrance blocks with various different widths of opening, which need to be altered depending on the time of year and the temperatures. With OMFs our hives have plenty of ventilation, so don't need a large entrance. I now have all my hives with a relatively small opening all year round. (See Fig. 2.10.)

The crown board goes on the top of the hive and is generally supplied with two elongated holes for use with 'bee escapes', but more about them later.

Lastly, there is the roof, which is normally fitted with either two or four ventilation slots. The slots are covered with mesh, but in many old roofs the mesh may be missing. These vents are supposed to aid ventilation but left to their own devices the bees would propolise any ventilation at the top of the hive. If these vents do not have the correct mesh cover, they will allow robbing by other bees or wasps.

I can't remember how many times I've seen a hive with a large gap between frames, or a frame missing completely, where the bees have built a large slab of wild comb. Often in the brood box this comb is full of brood and sometimes has the queen on it because it is new comb. If for any reason you have a gap between frames, push them all up together and put in a dummy board.

It took me many years of beekeeping to understand the importance of the dummy board. Eleven frames in a National hive with Hoffman spacing leave a gap at the end which can be filled with a further frame. Unfortunately, the frames are then so tight that it is almost impossible to remove the first frame to do an inspection. I had broken a few frames while doing this until I discovered dummy boards. The dummy board is much thinner than a frame so is easier to remove (or at least it should be). My first attempt at a homemade dummy board was too tight a fit into the brood box, as a result the bees glued the board in place with propolis. The board needs a bee space at the sides to make it work properly. I have

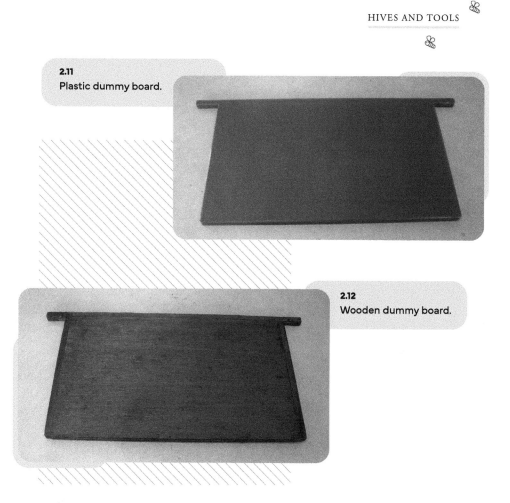

2.11
Plastic dummy board.

2.12
Wooden dummy board.

seen an empty frame, with a thin piece of plywood instead of foundation, used as a dummy board, but this doesn't really fit the bill either, as it is too thick and is just as difficult to remove as a normal frame. Plastic dummy boards are now available for all sizes of hive. They don't look much different to the wooden ones, but the plastic is more durable and you can't break off the end lugs.

The next consideration is the hive stand. Ideally, we should be able to inspect the brood box without having to bend; similarly, we should be able to lift off the topmost super easily. It is amazing how many beekeepers suffer from bad backs, due to bending and awkwardly lifting hive parts. I've come to the conclusion that for me the height of the stand should be about 40 cm (15″), and that means the WBC hive whose legs

2.13
Double hive stand to accommodate two hives.

are only about 15 cm (6″) is too low for me. The suppliers of bee suits come to the rescue here as a good suit has waterproof knees, which allow you to kneel to do your inspections. I've seen car tyres used to keep hives off the ground, but they are very low. I like to keep my hives on double stands, about 1.5 m (5 ft) long as this accommodates two hives and has room for another if making an artificial swarm.

I recall visiting one beekeeper whose hives were all located on a bank around the garden. While the hives were each on a stand on a concrete slab, the only flat area to stand was directly in front of the hive. In order to open the hive without being directly in the flight path, one had to stand at the side, on the slope with one foot up and one down. I shouldn't have really been surprised that the hives gave the appearance of being left alone and seldom opened.

Ideally, you should be able to get all round the hive, but the important thing is to be able to stand so that you are square on to the frames. All lifting of frames is then done without twisting.

What other equipment do we need? First, you need a smoker of sufficient size. My first smoker was an old one, which was quite small. It would barely stay alight long enough to inspect more than one hive. I recommend you spend as much as you can afford to get a good, big smoker. Find a fuel that works for you and stays alight, and you should have no problem. Some beekeepers use rolled up cardboard, as a fuel, but you have to be careful as many types of cardboard are treated with fire retardant. The smoke can be quite pungent. I use wood shavings, but they can burn quite hot and then your smoker becomes a blowlamp! Hay can be quite a good fuel, but burns very fast. You can often collect wood chips from the side of the road where trees have been cut to avoid overhead cables. Left for 12 months they are ideal smoker fuel. As a guide, if you find the smoke is strong and acrid to you then so will be bees. Try to get a nice sweet-smelling smoke. I know one beekeeper who uses cardboard, but has sprigs of lavender or rosemary rolled in.

I use a blowlamp to light the smoker, having spent many wasted hours trying to light up using matches and bits of paper.

Always light your smoker before putting on your veil. The material the veil is made from can be melted by heat, so sparks from lighting the smoker can burn holes in the veil. A friend of mine asked me to help him one day and he got fully dressed up before lighting his smoker. The chosen fuel was rolled up cardboard (not my favourite) which he lit and then blew on before putting it into the smoker. When I looked, he had burnt a large hole right in the middle of his veil, which of course he couldn't see!

When you've finished your inspections, stuff a twist of green grass into the end of the smoker and it will go out, leaving partly burnt fuel ready for next time. Don't put it in your car like that though unless you want the car to smell of smoke, and the smoker itself is still hot so will melt anything it touches. Similarly, don't put it away in the shed until you know it is out! I used to have an open box fixed in the back of my pickup and the smoker just went in there at the end of a visit to the apiary. If

2.14
Hey, you're on fire!

the bit of grass falls out en route, the smoker will just smoke away quite harmlessly. You will of course get some odd looks when stopped at traffic lights with smoke billowing out from under the cover! Once or twice I've been told that I'm on fire!

Hive tools come in an assortment of shapes and sizes, and generally, they are well made. I do remember being issued with one as a seasonal bee inspector that looked good, but it was easily bent, as the quality of stainless steel wasn't quite good enough. If you have an out apiary, I suggest you need a spare hive tool as I've mislaid one on a number of occasions, only to find it on a subsequent visit lying in the long grass. To keep your hive tool clean put it in a strong solution of washing soda. It will keep the tool free of propolis and kill pathogens.

Although not an essential piece of beekeeping equipment, it is necessary to keep the grass around the hives reasonably short, so I use a strimmer to keep the grass under control. It isn't popular with the bees, but it is quick. You could use other hand tools, but it really depends on how much time you have. It is better to do this type of maintenance in the evening when there is less activity in the apiary.

One essential piece of equipment, if you normally use them, is reading glasses. You need to wear them when inspecting your bees if you are going to stand any chance of seeing eggs. It is also important to remember to put them on before zipping up your veil. Once you are all zipped up and in the middle of a hive inspection, you can't put them on. While bending over they can of course fall off, but at least they are still inside the veil and can be put back on without having to unzip.

The other essential bit of beekeeping equipment, seldom mentioned in bee books, is the shed, or indeed sheds. I soon outgrew the space at the back of the garage, and needed somewhere to put all the spare bits of equipment when not in use. The more bees you have the bigger the shed you need, in which to keep all the essential, and sometimes non-essential, gadgets that you acquire over the years. For many years, I have attended the annual BBKA Spring Convention. As well as meeting old friends and listening to speakers on a wide variety of topics, there is a big trade hall. There are many bargains to be had, and there are always new items that every beekeeper should have. Over the years, I've bought lots of interesting although ultimately unnecessary 'essential' items, only for them to be used once and then put on the shelf. Of course, all this means you need a shed, or in my case, two sheds! One piece of advice on what to store in the shed – don't use it to store your bee suit. It is cold and damp in the winter and your suit will end up with mildew. Your gauntlets, which you rolled up together and put on the shelf, will end up as part of a mouse nest! The shed needs to be big enough to be used as a workshop, so will need a workbench and ideally an electricity supply and lighting.

The final 'must have' piece of kit if you have more than just a few hives, and have multiple apiaries is a four-wheel-drive pickup. A bit extravagant perhaps and may be not essential but it made my beekeeping so much less hassle. Using the family car for transporting hives and equipment is not perhaps the best idea!

3

Frames

Fundamental to beekeeping in a modern hive is the frame. I briefly touched on frames in the last chapter, but the things that can go wrong with them really merit a separate chapter.

First, it is necessary to put the frame together correctly. Most beekeepers get it right, but others sadly fail. One day I visited a local beekeeper in my role as a bee inspector. Before we actually got down to business he presented me with a bent strip of metal, with the announcement that I would need this for our inspections.

3.1
The mystery hive tool!

Bearing in mind that I had been a seasonal bee inspector for a couple of years and had probably opened over 1000 hives in that time, I had never found it necessary to use a tool such as the one I had been presented with. With hands already full with clipboard, smoker and bucket with hive tools, I slipped the new tool into one of the numerous pockets in my bee suit. Without further delay we set about our inspection of the dozen or so hives in the apiary. At the first hive I removed the roof, crown board, supers and queen excluder in the usual fashion and was about to remove the first comb from the brood box when I was told, 'You need to use that tool I gave you!' Not knowing quite what he meant, the beekeeper demonstrated its use, pushing it down between the combs, twisting it and lifting the frame from the bottom. Inspectors are generally a friendly bunch and not wishing to cause offence, I carried on lifting out the frames using his special tool. The inspection took longer than usual, but it was a fine day and the company good, but why did he insist on me using the tool? Suddenly the answer dawned on me and carefully looking at the frames I realised that there were no nails at all! If I had tried to lift the frames by the lugs on the top bar the whole thing would have fallen apart. I could tell that the hives didn't get regular inspection, or the frames replaced very often. I've since contemplated why there were no nails and I can only think that the beekeeper, frugal as we all are, must have saved money over the years by not buying nails. After all, frame nails are expensive – about £6.00 for 500 g!!!! That quantity of nails would be sufficient for about 250 frames, or 22 hives.

Once I did try gluing frames together using a hot melt glue gun. It was quick and worked well, but when I boiled the frames up to sterilise them before reuse they fell apart as the glue melted!

There is a correct way of assembling a frame and nailing it. The frame needs two nails to connect each side bar to the top bar. Opinions vary as to which way the bottom bars should be nailed. The normally preferred way is to nail the bar from the bottom, so that a bottom bar can be removed when re-waxing the frame for reuse. It is possible to insert a sheet of foundation without removing a bottom bar, so perhaps that argument is invalid. Bars nailed from the bottom on super frames have been known to be pulled off if the frames have been propolised together.

3.2
Correct positioning of frame nails.

Nailing the 'wedge' strip in place to fix the top of the foundation can cause difficulty. The frame nails are longer that the depth of the top bar, so if nails are driven vertically to fix the wedge strip they will protrude through the top of the bar. Once in a beginners' course a student managed to damage the tabletop by nailing in the wrong direction. Even if you manage to avoid nailing to the table, these protruding nails will tear your thin nitrile gloves and make scraping the top bar clean difficult. With frames that have wide top bars the frame nails are too short to nail the wedge bar in a horizontal direction, so it is necessary to drive them in at an angle.

I have often been asked for advice about buying first or seconds quality frames. From my experience, seconds are about 40% cheaper, but there will be some of the parts which due to knots and other faults won't be usable. I probably waste about 10% of the parts. Top bars with a large knot will warp in use and mess up the bee space. Taking all these factors

"Once in a beginners' course a student managed to damage the tabletop by nailing in the wrong direction."

3.3
Seconds frame parts that should best be discarded.

into account that still makes a saving of about 30%. Frames in supers are probably less critical, as long as they are strong enough to survive going through the extractor.

With brood frames it is important that everything is true, or bee space will not be correct. From what I've seen of other hives not everyone throws out as many bits as I do, and the misshapen frames seem to me to be a bit disappointing.

Now we come to the bewildering choice of frames. As with hives, the beekeeping suppliers' brochures have an extensive list of alternatives. Frames from different manufacturers can vary in dimensions, so this can sometimes compromise the bee space.

As well as 'discovering' bee space Langstroth also discovered the optimum spacing between frames in the hive, which he found to be 35 mm (1 3/8″), and his frames were made with this width.

A little later, the standard width spacing adopted in Britain was 36.5 mm (1 7/16″), which was achieved using metal spacers on the ends of the frames. We also had hives designed to take these frames with longer lugs to hold the spacers.

It is important to maintain the correct spacing in the brood box. Otherwise the bees will build comb as they want it, not how the beekeeper has provided for and intends.

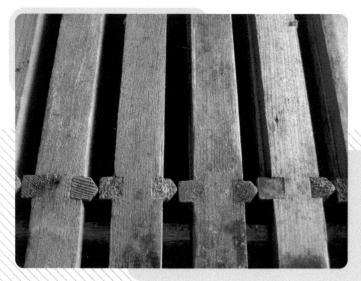

3.4
The frames in this picture are 'Hoffman' self-spacing frames with this 35 mm spacing.

3.5
Frames with plastic end spacers.

"With brood frames it is important that everything is true, or bee space will not be correct."

One problem with end spacers, which are made of plastic these days, is that they must be positioned correctly. If they don't abut properly the spacing is lost. Mixing different frame types will also lose you the correct spacing.

> "If they don't abut properly the spacing is lost."

3.6
Misplaced spacers, showing loss of correct bee space.

3.7
If we have a mix of different types of frame – self-spacing frames and ones with end spacers then the correct spacing will be lost.

For some years, I had been using 'brood and a half', which is a double brood box where the top box is a super. This makes a much bigger volume for the brood, but of course, there are twice as many frames. As an experiment, I tried using a 14×12 brood box. This is a National brood box with frames 12 inches deep rather than 8½ inches. The frames for these deep boxes are always self-spacing, Hoffman frames and came with 27 mm (1 1/16″) wide top bars, rather than the 22 mm (7/8″) wide

3.8
Hoffman frames with 27 mm wide top bars.

3.9
Hoffman frames with 22 mm wide top bars.

top bar that I had been used to. With the 35 mm (1 3/8″) wide spacing, this leaves an 8 mm (5/16″) gap between the top bars. As this is a bee space, the bees leave the gap between the top bars free of comb and propolis. With the narrower 22 mm (7/8″) wide top bars, the space is 12 mm(7/16″) and the bees often build comb between the bars. With the top bars free of comb, there is much less chance of trapping a bee when reassembling the hive. It took me a long time to discover this and I don't recall ever reading it in any book.

There is one proprietary hive on the market, which is made of plastic, modelled loosely on the Dartington long hive. Unfortunately, due to poor design or manufacturing tolerances, the box doesn't maintain the correct dimensions and allows the frames to be displaced sideways so that the self-spacing frames overlap, reducing the spacing to about 30 mm(1 3/16″). You only need to browse the internet these days to see some of the pickles new beekeepers get into, but strangely, they don't see what is wrong! What a potential trap this hive is for the new beekeeper, who seeing the advert thinks this is the ideal hive for his small back garden. It is also advertised as ideal for beekeepers who are disabled. It uses the large 14×12 frames, which are quite heavy when full. In addition, because of the design of the hive you stand at the side to do the inspection, so it means you have to twist. It is probably not ideal for anyone, let alone a beekeeper who is disabled.

The complications on spacing get even worse when we look at the options in the supers. Many years ago, it was discovered that once the frames had been fully drawn out, and extracted, the frames could be put back on the hive at an even larger space of 47 mm (1 7/8″) or even 50 mm (2″) using longer end spacers. A word of warning though. This can only be done with drawn-out comb. If frames with foundation are put into the super at this wide spacing, the bees will build comb between the frames.

There is one other spacing used in supers and that is Manley spacing. Manley was a commercial beekeeper in the 1930s and he discovered that the widest space for the bees to draw out foundation was 41 mm (1 5/8″). Unfortunately for the hobby beekeeper these frames, as with self-spacing Hoffman frames with wide top bars, do not fit into many smaller extractors.

"If frames with foundation
are put into the super at
this wide spacing, the bees
will build comb between
the frames."

I have seen hives where the beekeeper felt that the bees needed more space in the brood box so moved the queen excluder so that the hive was on brood and a half. Unfortunately, the super in question was on wide spaced frames and the bees built comb between the frames.

There is one other type of spacer worth noting and that is the castellated spacer. This spacer consists of a steel strip, which holds the ends of the frames. They are available as 9-, 10- or 11-frame arrangements for National boxes to give wide, Manley or narrow spacing. They are designed for use in supers, but some beekeepers, wrongly I believe, use them in the brood box. When opening a brood box, to avoid rolling bees when removing a frame, the preferred way is to lever the frame away from the adjacent one before lifting. With castellated spacers, it is impossible

3.11
Super with wide castellated spacers.

to do this. Many years ago, I had some plastic castellated spacers. They became brittle and broke. Happily, they are no longer available.

There remains one last type of spacing, and that is no spacers (sometimes referred to as finger spacing). I think the right way to describe this is the lazy beekeeper's spacing.

My preference for the brood box is to use Hoffman self-spacing frames. It is possible to buy plastic converter clips, which convert DN1 frames to Hoffman spacing, and I've tried two different types from different suppliers and they both have problems. First, they don't line up properly with normal Hoffman frames, so the spacing can be wrong. Second, because they go around either the side bar or the top bar they impinge on the bee space. I had some once that went round the side bar. This reduced the bee space and the bees propolised the gap and I couldn't get the frames out without getting a hive tool down the side to cut through the propolis.

I know many beekeepers who feel that the National (or WBC) brood box is not big enough for their particular strain of bee and have chosen

3.12
Hoffman converter clips.

to use brood and a half (a double brood box system where the top box is the size of a super). This is a perfectly reasonable system, but it does mean that when looking for the queen you have 22 frames to look at rather than 11. A particular problem I found when using this system is that the bees tend to build comb across the gaps between top and bottom boxes. When putting the hive back together after an inspection it is only too easy to squash bees, unless you take time to scrape off the top bars without decapitating any bees that happen to be looking up through the frames at the time.

For me the ideal solution was to change to 14×12 frames, but with my strain of bees I only need to have 9 or 10 frames in the brood box. This gives ample space for the brood, but means there is plenty of space for removal of dummy board and frames. In fact, with just 9 or 10 frames there is room for inspection without having to find somewhere to put the first frame removed just by moving the dummy board to the end of the brood box. Nine 14×12 frames in a brood box is still 14% increase in brood area compared to 11 frames in a National brood.

3.13
14×12 brood box with 10 frames and a dummy board.

The other advantage in using 14×12 frames is that the brood patch tends to be situated more centrally in the frame, and is round in shape, more like the natural shape of a brood patch, rather than a squashed oval which is seen on a National frame. The second advantage is that the brood patch is located lower in the frame, so that the arch of pollen all round the top of the brood area does not extend into the super. This means that the supers tend to be freer of pollen.

Good husbandry advice states it is best to regularly change combs, primarily to reduce the incidence of some bee diseases. Unfortunately, this advice rarely states how often is 'regularly'. From inspection of so many hives belonging to other beekeepers, I know that this piece of advice is often not heeded. One tell-tale of the age of a frame is when a lug breaks off. The obvious course of action is to remove the comb from the hive. I realise that to do so would mean that possibly 10% of the brood could be lost by the colony, but they will soon make good the loss and it could take some of the swarming pressure off the hive.

I recall seeing a frame in a hive at my association apiary that had a screw, but on a subsequent visit, probably a year later there were two screws in it as both lugs had been broken. These association hives were being used to teach beginners – never mind – just teach them bad practice! In fairness to the apiary manager, it is a thankless task looking after association hives, as it is effectively an out apiary that takes time that could be spent tending one's own hives.

> "One tell-tale of the age of a frame is when a lug breaks off."

3.14
The least obvious thing to do when a lug is broken is to insert a screw 'until I can get around to changing the frame'.

Apart from frame breakage, the main reason to change frames is to get rid of old, black brood comb. Every time a cell is used to raise a new bee, the pupae leave a papery cocoon in the cell, so over time the cell gets smaller and smaller. Pathogens are encapsulated in the comb. A good rule of thumb is to change the frames every 3 years. The first way to do this is to change three or four brood combs each year. The alternative is to change all the combs once every 3 years, and this is my preferred way. The advice suggested for changing just a few combs is to move the frames to be changed to the edge of the box at the end of the season. Then in the spring these frames will be unoccupied and can be removed. Whenever I tried this, on the majority of occasions, the winter cluster was not in the middle of the box, instead covering the frames that I wanted to remove. I ended up removing some younger combs from the other, unoccupied, end of the box instead. The inevitable result was that there were always some combs that were well beyond their best before date. The simplest way to change all the brood combs is to carry out a 'shook swarm', but many beekeepers find this a little bit too dramatic and prefer a 'Bailey frame change'. I'll describe these two methods in detail later. With the shook swarm method, you will be left with all the brood on the old combs, and although it might appear to be a great waste to not utilise all those developing bees there is a good reason for just destroying the combs brood and all. The sealed brood will contain most of the Varroa mites in the colony at that time, so this has the additional advantage of acting as a method of Varroa control. I'll cover that aspect later as well.

Having removed all the old combs, what can we do with them? Depending on how good the frames themselves are, they can have all the old comb removed and after sterilisation in boiling water and washing soda, then have fresh foundation fitted.

Wired foundation is useful as it makes fitting it into the frames much quicker than wiring the frames and fitting unwired foundation. It really depends how much time you have, but I prefer to wire my frames and then fit plain, unwired foundation, now that I have more time on my hands with fewer hives to look after.

The design of frames, with grooves in the side bars and two-piece bottom bars, is for use with wired foundation. The side grooves keep

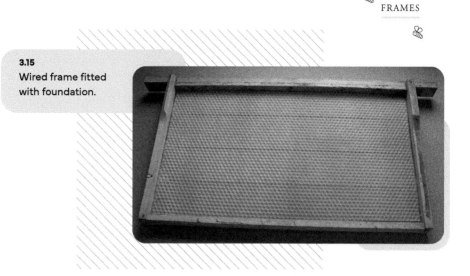

3.15
Wired frame fitted with foundation.

the edges of the foundation straight, and the two-piece bottom bars are supposedly to allow the wax to hang straight. I've seen frames where the bottom bars have been nailed together in the middle. That might keep the foundation in place, but make re-waxing impossible without first removing these nails.

With wired frames the side bars don't need grooves and one-piece bottom bars would be fine, in fact this is how most frames around the world are made. The foundation is trimmed so that it is slightly smaller than the distance between the side bars and short enough that it hangs just clear of the bottom bar. I think the bees like this arrangement as they often leave access ways between the comb and the side bars.

It used to be that making your own foundation needed expensive equipment, but there are now a number of cheap foundation moulds available. Foundation produced at home is generally thicker than commercially produced but, as it has been cast, it is more brittle than foundation produced by rolling. It does have advantages though. Apart from the satisfaction of making your own foundation, you know that your wax is free from harmful chemicals. Any chemicals that have been used in the hive for disease or pest control will be encapsulated in the wax, and commercially produced foundation is manufactured using beeswax from various sources, from both home and aboard, so we have no idea what harmful residues there are in the foundation. With a cheap foundation mould and a bit (no, sorry, a lot) of patience and practice it is possible to make acceptable quality foundation sheets. The whole process requires the wax to be at the correct temperature. Too hot and it will run out of the mould and the foundation will be too thin, too cool and the wax won't fully fill the matrix in the mould. If you get it wrong, and I certainly have done, you just re-melt the wax and try again. One lesson I did learn was that the work bench has to be perfectly horizontal, and my bench had a slight slope. Once I solved that issue the foundation sheets came out near perfectly.

As for wiring the frames, you do need a few, relatively cheap, tools. The holes through which the wire passes though the side bars need to have small brass eyelets inserted. You can get an eyelet punch which will make the job easier, but I found that it doesn't work on the wide part of Hoffman frames. It's almost as quick to drill the holes with an electric drill and the drive the eyelets in by hand.

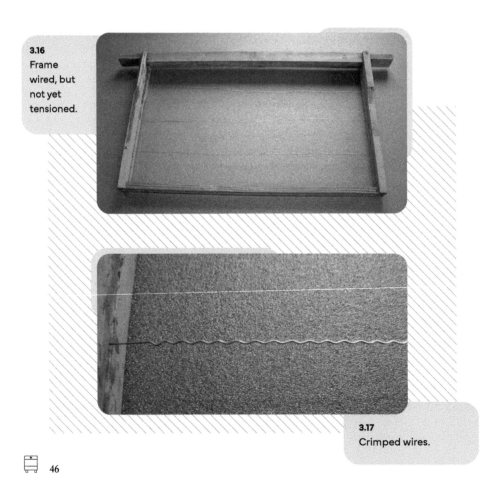

3.16
Frame wired, but not yet tensioned.

3.17
Crimped wires.

The wires need to be tight and although it is possible to just tension the wire by hand it is far easier to use a crimper.

The wires have to be embedded into the wax, and there are various ways available. I found the easiest is to use a 12 v electric current produced by a car battery charger.

Wire from different sources can cause problems. A good quality steel wire is best, but some stainless steel wires I have found to be too malleable and they are too soft to tension correctly, and bow when heated during embedding.

Beeswax foundation has a limited shelf life. If foundation is sealed in a plastic bag and stored in a cool dark place it remains usable for a few years. If it is left exposed to light and heat, it will become dry and loses its attractive smell and the bees tend to ignore it. There are too many other jobs to do in the busy beekeeping season to be putting new foundation

3.18
Embedding wire using a 12 v electric current.

into frames, so we tend to do this in the winter. It's always best to plan ahead and have fresh frames with foundation ready for the season, but if having prepared the frames, they don't get used then it might be best to melt the wax down and start again before next season.

4

Protective clothing

What about protective clothing? There are lots of suppliers and most of the clothing is reasonable quality. Some is very good, but some is cheap and shoddy. There is an expression 'buy cheap – buy twice' and this is especially true with protective clothing. There is nothing worse than a zip fastening failing when you need to deal with the bees. Many bee-keepers wear wellington boots, which was not my preferred choice when

4.1
Inspecting a hive with my little helper.

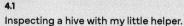

wearing them all day. For many years, I used to wear leather boots, but when I worked as an inspector my footwear had to be waterproof and washable, so that it could be sterilised, so I opted for 'muckers', which are short rubber boots used in stables. Now that I only have a few hives and I don't have to wear boots for long periods I do use short wellingtons, but only because the grass around my hives is normally on the long side and is often wet.

The choice of a full suit as against jacket and trousers is down to personal choice. A jacket is fine, but as it ages the elastic gets slack. It is easy for the jacket to ride up at the back when you bend over. Too easy I'm afraid, to expose the bottom of your back! There is a choice of hood/veil for your jacket or suit. Veils always used to be supported on a large round brimmed hat, often referred to now as a retro style or the more modern hood shape sometimes called 'fencing' style – I assume that refers to its appearance being similar to the face guards worn by fencers, not one that you wear while putting up a fence! The hood style doesn't offer as good protection to the head, and I've known bees sting through, particularly onto the top of my ears. Beekeepers who have bald heads should wear a hat under the hood as bees are often attracted to the heat on the top of the hood on a bee suit, so beware! The veil can also blow against your face, so the occasional sting on the end of your nose is to be expected. I do know beekeepers who wear a baseball cap which stops this happening.

"Veils always used to be supported on a large round brimmed hat, often referred to now as a retro style or the more modern hood shape sometimes called 'fencing' style – I assume that refers to its appearance being similar to the face guards worn by fencers, not one that you wear while putting up a fence!"

Most suits have multiple pockets. I've not managed to find a use for all of them yet. Sometimes it's easier to carry all your bits and pieces in a toolbox than to stuff your pockets with everything you might need.

Remember to zip everything up before going anywhere near the bees! I once was inspecting one of my out apiaries that had four hives in it. I completed three hives and was onto the last one when I noticed a bee inside my veil. I discovered that my veil was not zipped up. It is great having good-tempered bees! When you have finished your inspections, don't take your veil off until you are sure you have no bees on you. Bees will wait until you remove your veil before reminding you of their presence.

The legs of your suit should tuck into your wellingtons, as bees tend to walk up not down. For many years I wore leather boots and the elastic on the legs of my suit sealed well around the boot. As the suit got older the elastic became a bit 'tired' so didn't seal quite as well. One day while inspecting my bees, I felt a bee crawling up my leg. This can be a bit unnerving, particularly the higher up the leg it gets. What can you do about it? You are in the middle of a brood inspection and have a frame in both hands. The important thing I suppose is to keep calm, put the frame back in the hive and walk away from the hive, hoping all the time that the insect inside your trousers, which is still only just above the knee doesn't go much higher. You can't strip off, so the only course of action you have is to try to squash the bee. I don't like killing bees, but needs must. I squashed it, but it had the last laugh as it stung me on the top of my inner thigh. It could have been worse though, and that would have made for an interesting conversation at A&E!

Whatever your choice of PPE you need to keep it clean. When working as an inspector I always had two clean suits with me, so when visiting a beekeeper the suit I wore was freshly laundered. This was a requirement to avoid spread of disease but it does have other advantages. I once visited a beekeeper who advised that some of the hives were a bit bad tempered. I recall there were about 10 hives in the apiary and I worked my way round, opening each hive in turn and examining each brood comb, having shaken off all the bees from the comb. It's difficult not to upset the bees doing this, but they didn't seem too bad to me, in fact I'd come across other bees that were a lot worse. The beekeeper kept an eye on proceedings but was a few feet back from the hives. Almost at the end I said I didn't think they were too bad tempered, to which the reply from the beekeeper, who was covered in angry bees at this stage, was that she couldn't understand it, as the bees didn't appear to be taking any notice of me but she was covered in them. I had noticed that her bee suit looked as if it had not seen the inside of a washing machine for some time, if ever. What could I say, as I was there to inspect their bees not pass judgement on their laundry arrangements. Similar problems with aggression can be triggered by perfume or even sunscreen!

To wash your bee suit the manufacturer's instructions often say to wash the veil separately. I guess this is because the veil can be damaged in the machine. I know the way bee inspectors do it, which is to fold the veil inside the suit and zip up the front. One manufacturer recommends that the hood is detached and put inside the sleeve of the suit, this will protect it when on the spin cycle. Wash at the highest recommended temperature for the material, but at least 40°C, as a cool, 30°C wash doesn't destroy the pathogens that might be on the material. Put some washing soda in with the normal washing powder and this will help dissolve the propolis.

I have discovered that bees appear to know which parts of the human body are only covered by a single layer of cloth. In the height of summer it is tempting to wear tee shirt and shorts under your bee suit. From personal experience I now always wear long trousers and a long-sleeved shirt under my suit, so that I always have at least two layers between me and the bees when inspecting bees that are unknown to me. It does always depend on the bees.

Gloves are another item that should be clean. With so many colonies being less than ideal when it comes to temperament, there has been a trend towards gloves that are sting proof. There are a number of problems associated with thick sting proof gloves. These are made from either leather or a thick latex material. The latex type is washable, but because of the thickness of the material they can be a bit clumsy in use. They are also waterproof, so they don't breathe and your hands get sweaty when used for a long period. Leather gloves on the other hand are more

4.2
For some beekeepers I met in Cyprus a few years ago the normal attire was trousers and short sleeved shirt with no gloves or veil! Sunglasses are optional!

comfortable to wear, but can't be easily cleaned. I've tried all sorts of gloves over the years. I've washed leather gloves and found the resulting gloves to be stiff and difficult to use.

When I became a seasonal bee inspector, I was issued with thin nitrile gloves, which only provide minimal protection against stinging. That's perhaps not a problem with good-tempered bees, but some of the colonies that inspectors have to deal with can be less than ideal. Most stings that inspectors get are through the gloves, so they don't penetrate as deeply. Over the years the stings now have minimal effect, other than the initial pain. Each year inspectors are required to fill in a report on how many stings they receive in a season and what reaction they had to the stings. I normally reported about 600 stings but no adverse effects. The trick in the case of bad-tempered bees was to use two or even three pairs of gloves. They also make your hands sweat, so wearing them during long periods in summer can be uncomfortable. I discovered that a way of improving things is to wear a pair of thin cotton gloves underneath the nitrile gloves.

It is probably worth explaining why bee inspectors wear these thin nitrile gloves. Everything an inspector wears has to be clean, particularly gloves. Disposable nitrile gloves are the only way of ensuring that diseases aren't spread between apiaries. An advantage to the beekeeper when wearing these thin gloves is that you can feel what you are doing, it is even possible to feel bees walking over your hands. It is easy to move bees out of the way when reaching in to take hold of the frame lugs, thus avoiding unnecessary squashing of bees. When I stopped being an inspector, I continued to use nitrile gloves, but after a few years the long cuff gloves became unobtainable and only shorter gloves could be purchased. These don't allow the sleeves of your jacket to be tucked in sufficiently. I had already spoken to the inspector who had taken over from me, and he had the same problem, and now they are kitted out with gauntlets to cover the gap between glove and sleeve. This isn't an ideal solution as the point on the wrist where the gauntlet meets the glove is now a place where bees get accidentally caught, and when that happens, they do sting. You can get over this problem by pulling the gauntlet down over the wrist and then pulling the gloves over the gauntlet.

Sometimes the nitrile gloves can get torn by catching on a protruding nail, a rough edge or a piece of wire. Just put another glove on over the top of the torn one! Don't try to remove the glove and replace it, as your hands will be damp and the replacement glove won't pull on.

Some bees gather large amounts of propolis, and in the warmth of the hive this can sometimes be a sticky problem with fingers becoming covered in propolis. Impossible to remove from leather gloves, but easy with nitrile ones – I just wash my hands in the bucket of washing soda that I keep my hive tools in. Not possible with leather gloves though.

There is one advantage to washable/disposable gloves and that is to do with stings. When a bee stings it will leave the sting attached to the sting site, and that will continue to pump poison long after the bee has gone. It also leaves an alarm pheromone at the sting site even when the sting has been removed. This pheromone will remain for days, or even weeks, so the next time a glove that a bee has stung is used, it will have alarm pheromone on it which will attract other bees to sting. A beekeeper with dirty gloves and traces of old bee stings will be wafting alarm pheromone over the top of the hive – not really conducive to having calm bees. To avoid further stings, once a bee has stung on your glove, puff smoke over the area to mask the alarm pheromone.

If you look at pictures in most of the old books you will see that gloves are infrequently worn, not because these old beekeepers were braver than we are, but probably because the bees were better tempered. Beekeeping with bare hands is difficult initially, but if your bees are docile it soon becomes easier. It's a revelation when you first open a hive with bare hands and you realise for the first time how warm it is in a hive! There are some operations that require bare hands – picking up a queen for marking or clipping for example. Some of these things can be done with nitrile gloves on, but they have to be free of propolis, so one advantage of wearing gloves is that they can be removed if required and your fingers are free of sticky propolis. Putting gloves back on afterwards isn't as easy though, as invariably hands are damp from perspiration, so if you are planning on removing your gloves to pick up a queen have a towel with you so that you can dry your hands.

"It's a revelation when you first open a hive with bare hands and you realise for the first time how warm it is in a hive!"

5.

Moving bees

Beekeepers have been moving bees for hundreds of years. The Egyptians loaded basket hives onto small boats, which sailed along the Nile in search of blooms. Long before the time of our modern moveable frame hives, in the early 17th century settlers to the New World took bees from Europe to pollinate the European fruit trees that they had also taken there. These bees were referred to by the indigenous population as 'white man's flies'. It is not at all clear how the transportation was done, and it took a further 200 years before they were taken across land to reach the West Coast of the USA. The first bees to successfully arrive in Australia

5.1
'I told you we needed more straps!'

were carried on board a prison transportation ship in 1822, and the first bees to arrive in Tasmania were again moved on a prison transport ship in 1833. These bees were native British bees, and in fact one of the purest strains of *Apis mellifera mellifera* (AMM) is still to be found in Tasmania. There is very little information about how this transportation was done, but there are many reports of bees dying before they reached their destination during the seven and a half month voyage. Little did those early long distance transporters of bees know what long-term impact they would have on the environment. Now the movement of bees is much easier and is in the opposite direction. It was only a few years ago that these re-importations from the New World were stopped due to the risk of importation of small hive beetle. Might what was seen as progress 200 years ago be seen as a big mistake in hindsight?

All beekeepers, at some time, will find it necessary to move bees. It's a simple enough task, but easy to get wrong. The first time a new beekeeper has to move bees is when they acquire their very first colony. This is normally in the form of a five-frame nuc, but may be a full size hive. With any luck the nuc should have been prepared properly by the seller, ready for transportation.

Some of the old books refer to nucs being purchased by mail order and delivered by train. The beekeeper would collect the bees from the station. How things change!

In my case, the first colony of bees I bought was in an old WBC hive that was falling apart, although I didn't know that until I came to move it. I purchased a colony of bees together with a pile of old equipment from someone selling his deceased father's bees. He had no interest in them and it was all down to me to move them. I had read up on how to move bees, but that didn't tell me what to do when the bees started to come out of the gaps in the old hive. They needed to be moved about 10 miles (16 km), so it shouldn't have been much of a problem. I left the job until the evening, when the book said they would have finished flying. Armed with strips of foam and some straps I closed the hive entrance up and strapped the hive together. The hive was then lifted onto my trailer. Thank goodness I had a trailer, as I would have been in trouble if the hive had been in the car. Off I set with the bees on the short journey to my new apiary, in a friend's garden on the outskirts of the village. I arrived and it was almost dark. I did notice that there were lots of bees on the outside of the hive! All that was necessary now was to lift the hive out of the trailer and carry it about 50 m to the bottom of the garden. Wearing my newly purchased veil and a pair of overalls, trouser bottoms tucked into socks, I carried the hive to its resting place. By this time, I had bees on me and stinging my ankles. The hive was set down and the foam removed, although thinking about it later I'm sure the bees had plenty of other ways out of the hive than just through the door. At this point in almost total darkness I beat a hasty retreat, wondering if this idea of keeping bees had been a good idea in the first place.

I had purchased these bees in September, so they needed feeding to get them through the winter. The plan was to leave things as they were until the spring and then transfer the bees into a new hive. There was plenty of other beekeeping equipment that I had bought as part of the job lot, and the winter was spent sorting out and cleaning up enough of the hive parts to make a new hive. The bees made it through the winter, were put into their new hive in the spring and from there on I was a beekeeper.

Since then I've moved bees many times, and now have learnt, the hard way, how to do it with the minimum of hassle. The one rule you have to work to though, when moving bees is the 'rule of 3'. You move them more than 3 miles (5 km), or less than 3 ft (c.1 m). Why? Because bees will always return to the location of the hive, so if you move a hive say 6 ft (c.2 m) they will return to the old site and be confused, before trying to go into the nearest hive. When moving a distance, because bees generally forage in about 2 miles (3.2 km) radius of the hive, moving more than 3 miles (5 km) will eliminate the possibility of them coming back to the old location.

Some years after I started beekeeping, I was given four hives of bees and a lifetimes worth of accumulated beekeeping equipment by a relative of my wife in Derbyshire, as he was retiring from beekeeping. The whole thing needed three trips, each one 75 miles (120 km) and taking 2 hours. Moving the hives was the first trip. This time I was more prepared and had some help loading up. He had closed the bees up the night before and had put ventilated screens on the top of the hives. Suitably strapped up, the beehives were put into my trailer and off we set. The journey took us over the top of the Pennines and through Stockport, and then on to the M62 and M56 motorways to my home in North Wales. It was midday when we arrived in Stockport and the traffic was stop–start. Fortunately, no bees escaped from the hives, because we spent some minutes stationary by a queue waiting for the bus. It wasn't a hot day, but I was sweating buckets sitting in the car wearing a bee suit, worrying about escaping bees and people getting stung. All went well though and finally we arrived at the new apiary site where the four hives were to be placed. The hives were set up on the previously prepared hive stands, the foam taken out of the entrances and the roofs replaced. We stood back expecting the bees to come rushing out, but after a short time, only a few ventured out to have a look at their new surroundings. These colonies were AMM bees that had been bred as part of a bee breeding programme and it was my experience with these dark, docile bees that encouraged me to change all my bees to this particular strain.

Not every move went so well. The first time I decided to take bees up to the heather moors was an interesting experience. We live about 1.5 miles (2.5 km) from the nearest heather, so the options were to take bees from an out apiary to the nearest moorland area, or move bees from home to a location further away. On this first occasion I had bees at a site a few miles away, so it was decided to move these to the moorland site near home. The bees were prepared for the move the night before, and strapped up using a single strap on each hive.

In the morning, we collected the hives with the trailer, and set off. No real problems on the way, until we got to the old drovers' road over the mountain, which was rutted and uneven. The hives rattled about and I noticed in the rear view mirror that some bees were sitting on the top corner of one of the hives. We came to a gate and my wife got out to open it, and seeing the bees put her veil on. I drove through and before she could get back into the car a runner came along the track towards us.

I never knew quite what happened next, but I noticed in the mirror that as he got alongside the trailer he suddenly broke into a sprint. I

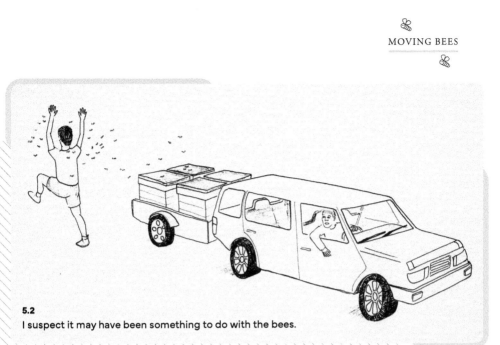

5.2
I suspect it may have been something to do with the bees.

like to think that it was part of his training routine and that he always sped up at that point. I suspect it may have been something to do with the bees. What had happened with the hives of course was that with just a single strap and all the jostling about on the rough track, one of the hives had twisted and allowed some bees to get out. For the last half a mile my wife jogged alongside, holding the hive together. I made a mental note to use two straps per hive from now on.

5.3
Hive with only one strap.

On another occasion, I purchased some hives from a beekeeper who was retiring. The hives were tucked away behind a small thicket. The weather was quite cool, so I decided to get there early, before the bees were out and about. As planned, when I arrived the bees were nice and quiet and hadn't started flying. The entrances were all closed up using foam and the hives all strapped together with two straps this time. So far so good, but fighting my way out of the thicket carrying the first hive I caught the corner of the floor on a branch. It moved sufficiently for bees to come pouring out. I put the hive down and twisted it back straight, but the harm had already been done. The hives were all loaded onto the pickup and the rest of the move was completed without further problems, with a few of the bees riding on the outside of the hive.

I made another mental note that if using two straps then they have to be parallel not crossed over and then the hive can't twist. Why had I never read that in a book?

5.4
Hive strapped up with two straps – the wrong way!

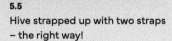

5.5
Hive strapped up with two straps – the right way!

Whenever we intend to move bees we should also have a backup plan. I once was planning to collect some hives in late June that had been on the oil seed rape. The honey had been removed previously but the hives needed to be moved from the site. They were already strapped up, and just needed the entrance closing. The plan had been to go in the evening, about 9.00 pm, when the bees had stopped flying and then take them back to the home apiary that evening. It was a warm summer's night and at 10.30 pm, the bees were still flying. It was almost dark by the time all the foragers were back in the hives. It always helps if you have a torch, but at the time that wasn't on my list of essential equipment. There was just enough light to see to close up the entrances and get the hives onto the trailer. Unloading them would have to wait until daylight.

I did hear of some beekeepers who were taking bees to the heather moor. They were delayed by roadworks while in transit and arrived at the moor so late that it was dark. They couldn't even find the stands that they had set up and ended up camping out under the stars until the morning.

When moving colonies to temporary sites, don't forget that you have to take the hives home after the crop finishes. I once took four hives to some oil seed rape, and the farmer asked me to put them by a hedge between the two fields of rape. There was a track across the field, so access was quite good. I was able to drive my car and trailer across the field to unload the hives. Unfortunately, when the crop had finished flowering the oil seed rape was about 6 ft (2 m) high and the track had all but disappeared. I never realised that oil seed rape grew that high. Carrying the four hives back through the field of head high rape, uphill, was to say the least, difficult!

The condition of the hives can also present problems when moving bees. Most wooden hives are made out of cedar, but some of my open mesh floors were only softwood. Over the years these floors often start to rot, or the mesh becomes loose. This isn't a problem until you want to move them. All looks well until you put a strap around and tighten it up only to have the floor partially fall apart. Alternatively, it may all look OK, but during transit bees can get out. This in itself isn't a problem if they are on the back of a pickup or in a trailer, unless you are driving through a built-up area of course. It's worth checking the condition of the hive parts before contemplating a move. Much of my new equipment has been purchased as seconds and occasionally knot holes in the timber stop the hive being absolutely bee proof.

The general advice when moving bees is to use a travelling screen, which is a mesh board, to replace the crown board on the hive. Without the roof in place this allows plenty of ventilation. When travelling any distance with hives in a trailer or on the back of a pickup they will have plenty of ventilation. If you are moving bees in a car there are other necessary precautions.

My first disaster moving bees in the car was when I moved four hives to the heather moor about half an hour's journey away. The hives were prepared in the usual way with travelling screens and the entrances plugged with foam. The hives were loaded into the back of my hatchback, as at the time, this was easier than putting them in a trailer, with each hive sitting in its own roof. The original plan was to take the bees straight up to the moor, where hive stands had already been set up, but as we thought it would be dark when we got there, we decided to leave it until the morning before driving up to the moors. It was a lovely sunny morning and there was plenty of time, so we had breakfast first. When we opened the car, which had been parked on the drive all night, it was like opening an oven door. I had no idea how hot a hive could get. We poured water into the tops of the hives, but even so, one of them had got so hot that the combs had started to melt and honey was running out of the bottom. The hives were taken and set up

"The lesson I suppose was that even though the hives had plenty of ventilation the car didn't. We hear lots of warnings about leaving dogs in hot cars, and the same is true of beehives."

on the moor, but that one hive was dead. The lesson I suppose was that even though the hives had plenty of ventilation the car didn't. We hear lots of warnings about leaving dogs in hot cars, and the same is true of beehives.

I once spent just over a year working on a construction site near Doncaster, about 2.5 hours drive away, and there was plenty of space on site for a few hives. One spring morning I set off from home with a nuc that had been prepared for transportation the night before and took it over to its new temporary home on site. Over the next few months this built up into a full hive and was split into two colonies and two local swarms were collected. By the end of summer there were four hives that needed to be brought home. The hives were closed up one evening to be transported the following day. At about 4.00 pm they were put into the back of my hatchback car and we set off for home. I kept my bee suit on for the journey just in case, not wearing the veil of course. In spite of pouring water into the top of the hives, they produced so much heat that I had to drive with the windows open (this was prior to the days of air conditioning in cars), to keep me cool as much as for the bees. I should have remembered how much heat one hive can produce, but certainly four was like having the car heater on full blast, added to which I had my bee suit on. All went well, but driving with bees in the car is quite stressful. One wouldn't want to be in a road traffic accident with a car full of hives as I doubt the emergency services with all their equipment have the right PPE to deal with bees.

Some years ago, I was working on a construction site in Herefordshire about 100 miles (160 km) from home and took eight hives to a field of borage near to where I was staying. The move to the site was fairly straightforward, closing the hives up at night and loading them up on the trailer to be moved first thing in the morning. After an almost 3 hour journey the hives were in place by 9.00 am. The return journey was in two parts, first, the honey supers were removed and transported home for extraction and the following week the hives were transported home. They were closed up with travel screens on in the evening and loaded onto the trailer ready to be moved the following day. I couldn't move them until the evening, as I was working. The weather was hot and sunny, and the hives had no shade, so twice during the day they had water tipped through the travelling screen to help keep them cool. I set off for home about 5.00 pm and it was still very hot. I had to make a stop for fuel on the way back, and the bees were given more water. I got some odd looks from other customers at the garage as having put diesel in the car I poured water into the hives on the trailer. It was about 8.00 pm when we finally got home, the bees were put back on the hive stands and the entrance blocks removed. No problems this time, no escaping bees and no overheating – after many years of trying, at last, I must have got it right.

Most of my beekeeping has been on my own, but if at all possible I recommend getting help when moving bees. Carrying hives across rough moorland is not easy on your own. My first attempt at setting up bees on the heather moor on my own was eventful. When carrying a hive, you can't see where you are putting your feet and moorland is uneven with lots of hazards. When you trip and fall over into a muddy bog you really test the hive strapping, particularly if you have only used one strap. The bees wouldn't have enjoyed the journey so far, and you have just dropped them and given them even more of a shaking up. By the time you have picked yourself up and checked whether you have done any serious damage to yourself (and had a quick look round to see if anyone has witnessed you falling over), the bees will have found their way out of a corner where the hive has twisted in the impact with the ground. At this point you are covered in mud and wet through. You have to get the hive over a fence and set up on a stand, and then if that isn't bad enough, you still have three more hives to carry the 200 m across the moors. Any future moves to the moors were always done with a helper, and we rigged up a hive carrier with two long poles and a couple of straps – which we called the sedan chair for hives.

If anyone ever tells you that your heather honey is too expensive, just remember all the trials and tribulations getting bees onto, and then off, the moors. Not to mention the problems of extraction, but that's something for later in the book.

Another growing problem with bees in isolated locations such as heather moors is theft. A hive with a full heather crop is worth a lot of money and ever since I have kept bees there have been reports of hives having been stolen.

I first came across an account of this in the early 1980s. Some local beekeepers had apparently been helping each other with hives up on the nearby heather moors. One Saturday morning they set off to check on the hives and passed a lorry coming in the opposite direction loaded with beehives. They thought that unusual as they had never seen a lorry loaded with hives before. When they got to where their hives should have been they found they had disappeared. It dawned on them that they had probably been stolen and transported on the lorry they had seen earlier. It was reported to the police and later the same day a lorry was apparently found in a motorway service area heading towards the south coast.

There are numerous reports of hives being stolen, and sadly this has to be by beekeepers who have the correct equipment and enough knowledge to know what they are doing. From what I have read, theft of bees in other countries takes place on a commercial scale.

So far, all of these stories have concerned bees being moved a substantial distance. Things are much easier when moving shorter distances between apiaries. All my hives now have open mesh floors as part of my Varroa control measures, so for moves that can be completed within 15–30 minutes I don't bother with travelling screens. I just strap everything up with the roof in place and close up the entrance, and move them in the evening. Provided the roof is bee proof and the hive is sound the move generally goes well. I use a plastic tray under the hive to protect the carpet in the back of the car and I'll write about these trays later. (For a summary of how to move a distance see Appendix 2.)

In the UK we are taught all sorts of precautions when moving bees, but having spoken to beekeepers who have worked with commercial companies in New Zealand I do wonder whether we make rather heavy weather of things. They don't use travelling screens and apparently don't even close the entrances. The hives are just loaded onto a truck and a fine mesh net put over the whole thing. In the USA hives are moved in their thousands every year for pollination, particularly for the Californian almond crop. Often one hears of a lorry carrying bees being involved in a crash, with beehives strewn across the road.

What about moving hives within the apiary? The rule here is to move them less than 3 ft at a time. Normally when you have a swarm from your own apiary they hang up in a nearby tree or hedge (usually next door I might add). The swarm can be collected and allowed to settle into the collection box until the evening and then re-housed into a hive. If it is a prime swarm, that is, one with the original old queen, they normally settle into the new hive very easily. This does not always happen with a cast, or second swarm. In this case the queen is a virgin and the swarm is far less easily satisfied with the beekeeper's choice of new home. I once had a cast that collected on a branch of a tree in my garden – so far so good because that was a first as they normally go next door. I shook them into my swarm collecting box and waited for them to settle. Out they came and went back onto the branch. Plan B was to shake them into the box and strap the box on top of the branch – they were sure to go in then! Some did and then came out. In the evening I shook them all back into the box and they appeared to stay, so I put the box on a hive stand in the apiary about 20 m away with the intention of putting them into a hive the following day. Next morning the box was empty. No sign of the bees! I looked at the branch in the tree and there were no bees there other than a few flying about, but glancing down, there was the swarm in a heap in the grass. Now we are on to Plan C. I made up a nuc with four frames of old but clean combs and a frame of honey. I removed the central frames and scooped handfuls of bees into the nuc box and then replaced all the combs. This box was then put with the entrance right next to the pile of bees. Over the next few hours all the bees went into the box and not wanting a repeat absconding of the previous day the box was left for two more days for the bees to settle in.

That had solved one problem in that I had collected the swarm, but they were 20 m away from where I wanted them and there were two fences in the way. Observing the 3 ft rule it would take many days to get these bees in place. First move was to lift the box into a wheelbarrow. Next day the barrow was moved 3 ft. This was repeated every day for 3 weeks until the bees were through the gate into the apiary and then left to settle down. During this time, as the hive was moved each day the bees would fly down the path and then on to the nuc box. It was almost like the path was an approach strip to the runway. A couple of days later I noticed that the bees were bringing in pollen and I had a look in and the queen had already started to lay eggs. Success, but it would have been easier if I hadn't let them swarm in the first place.

I've visited apiaries where the beekeeper has the odd hive away from the rest, because that was where the swarm was put into a hive, and they've never got round to moving it nearer to the other hives.

One of the manipulations that beekeepers often need to do is to unite two hives. There are various reasons why we would want to do this, and different ideas on the best way of actually doing the uniting. However, the first thing we would need to ensure is that the two hives to be united are situated next to each other. You can't unite two hives that are at different ends of the apiary, as the foraging bees will return to the original site of the hive. Having stated that rather obvious point, I have known even experienced beekeepers who fail to observe that principal. One or other of the hives will have to be moved so that they are adjacent to the other. This can be a time-consuming process, so very often beekeepers don't bother.

There is one time of the year when all these rules about how far you can move a hive no longer apply. During the winter months the bees are dormant and very little flying takes place. At this time hives can easily be moved any distance. Winter is the time of year if you want to move your apiary from one place to another a few hundred metres away. All you need to do is gently move the hives to the new position. If this is within the apiary you don't even need to strap the hive up. Bees can be moved long distances in the winter, but this is not recommended. Once the bees are in their tight winter cluster we don't want to disturb them any more than is necessary.

6

Records, planning and inspections

When I started out with my bees, I had no real plan, I just thought that it was an interesting rural hobby. I might even produce some honey that would taste as good at that we got from the school bees all those years ago. At some stage a vague plan emerged about how things might

6.1
'What are we going to do?' – 'Don't know, what are we going to do?'

progress. For many years I had just a few hives, but after early retirement I decided to expand to provide an extra income. It was really a case of start small and make mistakes on a small scale, before going on to do the same on a big scale!

I guess that with most beginners their plan is just to keep bees, and there might be some vague idea about producing honey. Producing honey in any quantity is something many beginners fail to achieve in their early years. Much of their time and efforts is spent trying to stop the bees from swarming, keeping the bees healthy (particularly dealing with Varroa), getting the colony through the winter and generally trying to work out what on earth is going on in the hive.

The first question that the beginner needs to ask themselves is whether they want to keep a very small number of hives, say one or two, or are they intent on keeping more than 10. I was once asked by a would-be beekeeper on one of my beginners' courses if taking on 12 hives from a retiring beekeeper was feasible. I forget my reply, but whatever it was it didn't deter the lady in question from taking on the hives. It takes time to realise what it is that you have let yourself in for when you take up bee-keeping. Over-enthusiastic new beekeepers can soon realise they have bitten off more that they can chew.

In my first few years of beekeeping I kept no records. With only one or two hives I couldn't see the point, as I thought I could remember what was going on. I was wrong, however, as I easily confused what was happening in different hives. The simple reason for keeping records is that when we look in a hive we see only a snapshot of what is going on at that time. What we see is the result of what has been going on over the past 1 or 2 weeks. For example, when we see signs of queen cell production, this has been triggered by conditions over at least the previous 2 weeks – perhaps due to overcrowded conditions, confinement in the hive due to adverse weather conditions or a change in forage availability. In those early beekeeping years, I lost swarms and produced only a little honey.

I know from my years as a seasonal bee inspector, that many bee-keepers span the range of keeping no records to keeping such detailed information that I wonder how useful those records would be. Perhaps some find record cards too difficult. Others, perhaps with a scientific background, record things in meticulous detail. I think the important thing is to ensure that if another beekeeper were to open a particular hive he or she would from the records be able to understand what had gone on over the previous weeks and what changes had occurred. I've often been asked by local beekeeping associations to take a lead in an apiary meeting. With no records, all one can do is to try to observe what is happening and try to guess what course of action, if any, you need to take. Everyone is disappointed, or perhaps amused that this 'expert' that has come along, can't find the queen, which has hardly ever been marked (marking the queen was always the other task that one is asked to do at these meetings).

I developed my own system of records, and the card acts as an aide memoire when looking through a hive. The requirements when inspecting a hive are:

- Has the colony sufficient room?
- Is the queen present (and laying the expected quantity of eggs)?
- (Early in the season) Is the colony building up as fast as other colonies?
- (Mid-season) Are there any queen cells present?
- Are there any signs of disease or abnormality?
- Does the colony have sufficient stores to last until the next inspection?

What we need, therefore, as a record card, is a simplified way of recording all these things without making the task too onerous.

Some beekeepers keep their records in a notebook, others on a card stored under the roof of the hive. There are two problems that we need to address here. First, bees stick everything together with propolis and this gets all over your fingers. Your pen will get sticky and so will your notebook. In these days of smart devices, we could use a phone or tablet in the apiary, but it too would soon become covered in propolis. The second problem is that any object in the hive that the bees don't want will be chewed up and removed unless it is protected. My first attempt at leaving a record card on top of the crown board was soon discontinued – by the bees that is, not by me. They simply chewed up the card and left it as a pile of papier-mâché on the ground outside the entrance to the hive.

For a few years I looked after the apiary for our local association, and we held regular Saturday afternoon meetings during summer months. One week, I knew I would be unable to attend that weekend, so I inspected all the hives mid-week to ensure that all would be well for the coming meeting. One of the hives had a queen, but she had not yet mated. In order that this particular hive should be left alone I put a note on a piece of card on top of the crown board suggesting that this hive should not be opened at this time as there was a virgin queen there. Following on from the apiary meeting, I had a phone call from the person who had led the apiary meeting. All had gone well, except that one of the hives looked to be queenless. I asked about the note I had left and was told, 'There was no note!' I visited the apiary a few days later and sure enough there was my note in tiny bits outside the hive.

One can put the record card inside a plastic envelope, which will protect it or alternatively it can be laminated and written on using a suitable pen that will write on plastic. Generally, the system of laminated cards works well, although on one occasion on a rather exposed apiary site that I used for a short period, the wind blew the card away and I never did find it.

My record card (see Appendix 3) should be regarded only as a suggestion and can be modified, adapted or even rejected to suit one's individual needs. When selecting colonies to use for expansion I wanted to use only the best tempered ones, so more columns were added to the card. Each hive was then given a score for good temper. The bees are always better behaved when the weather is warm and sunny, so a further column was needed to record the weather.

The most important record is often what we write in the comments column, so for this reason this is the biggest column.

The first question to answer is does the colony have enough space. Here there is some disagreement between beekeepers as to whether a colony needs more space for brood than a single brood box. If you have near native bees then a single National brood box should be large enough. Those beekeepers who are trying to manage other non-native strains or hybrids may need larger brood boxes and may have to use brood and a half (a two-box system utilising a normal brood box with a super on top) or double brood boxes.

In answering this particular question, we need not consider the brood box size. I assume that we have the right size for our particular strain of bee. The question we need to address is whether the colony has enough space for the honey. Don't forget that the bees need space to process the honey and not just to store it. We need to give enough space for the bees, not just space for honey. My rule of thumb here is we need to put on another box (super) when the bees are covering about 80% of the frames in the topmost box. It is also better to give them more space than they need than too little. If they need additional supers, I record this in the comments column on the card.

Once we have a record system in place things suddenly become clearer. A quick glance at the card before opening the hive will give us an insight into what we might see. The card should tell us if the queen is marked, and what colour the mark is.

Don't always expect there to be only one queen. I know there is normally only one queen in a colony of bees, and that is what the books tell us, but I think there are more instances than we might expect of perfect supersedure, where there is both mother and daughter in the hive. A few years ago, I had a hive that superseded. There was just one queen cell, and the laying pattern of the 2-year-old queen was looking slightly uneven. The new queen emerged and about 3 weeks later the brood pattern suddenly expanded. The old queen, who was marked red, was nowhere to be seen, although the mark had become worn away so was less obvious and difficult to see. The new queen was marked with a blue mark. The following year the colony expanded very rapidly, and I carried out an artificial swarm when I first saw signs of queen cells. I found the queen, with her blue mark, and transferred her on the frame she was on into the new hive as normal and relocated the old hive to just a few feet away, leaving just one open queen cell. A week later I inspected both hives and found to my amazement that there were fresh eggs in the old part of the split hive and no queen cells. A thorough inspection revealed the old queen, still marked with a very faint red spot, still laying eggs! These two queens had both been in the same hive for 8 months.

I have to ask myself how often this happens. I've noticed on occasions when having done a split on a hive that an unmarked queen is laying in what was expected to be the queenless half. The lesson here is look for a queen, and not necessarily a marked queen. Sometimes marks wear off, just to confuse matters.

The next thing to take note of during our inspection is what change there has been in the brood since the previous visit. In the early part of the season we would expect an expansion week by week, but if this is not happening, we need to investigate why this might be. Look at the ratio of eggs to larvae to sealed brood. In a stable situation it should be 1 : 2 : 4. If the ratio of eggs is higher then the colony is expanding; the converse is true if there are fewer eggs.

We may observe queen cups (the start of queen cells proper) and should check to see if any of them has an egg in there. If there is an egg we need to consider if we should do something about it. An egg could be just laid or 3 days old. Some beekeepers find it difficult to see eggs, let alone guess how old they are. Apparently a freshly laid egg stands straight up from the bottom of the cell and will gradually lie down over the next 3 days. I'm afraid that I never been able to make that particular judgement. In the case of a queen cell the egg is always hanging vertically down, so it is impossible to judge the age. This egg will turn into a larva and the cell is then sealed on day 9 after the egg being laid or 5.5 days after the egg hatches. On balance therefore most eggs will advance to the sealed cell stage by the next 7-day inspection. We need to make a decision about what to do. I normally at this stage check through and see how many other queen cells there are containing eggs and also check through thoroughly and find the queen. If we can't find the queen, the task just becomes more difficult. Do we assume she is still there, or has she gone for some reason? Unfortunately, only experience will be able to help here. If we look at the brood pattern we should be able to assess whether the number of eggs in the brood generally is what we would expect (assuming the beekeeper can see eggs). If there are plenty of eggs then assume the queen is still there, even if you can't find her. At least you know she has been there during the last 3 days. Cut off all the queen cells you can find. This will ensure you have at least 1 week before the colony will be in a condition when it might swarm. This will also allow you the time to get together all the equipment needed to do your swarm control. Provided there are eggs, even if we have misread the situation and the queen is no longer there, cutting out queen cells is OK. Importantly record everything you have done.

Beware when looking for queen cells that the bees will often obscure the cells by clustering over them, so move the bees off all the nooks and crannies where queen cells might be.

The card will help to plan all these things as you will have a record of what has been seen previously. I don't specifically look for the queen at every inspection, but if she is seen then I tick the box on the record card. What I really want to see is whether there is evidence of her being there. If there are eggs then she must have been in the hive during the previous 3 days. If there are plenty of eggs then she is almost certainly still in residence.

6.2
Two queen cells obscured by the bees!

Our next point to address is disease. I make a point of doing a full inspection for disease once a year. To do that we need to inspect the brood combs. That can't be done when the frame is covered with bees, so you have to shake all the bees back into the brood box so that you can look for diseases of the brood. I'll cover this aspect in Chapter 8. One particular thing we need to check on regularly is Varroa mites. I hadn't had to worry about this problem until 1996, when the first Varroa mite was found in one of my hives. At that point beekeeping became more difficult. Generally, most beekeepers have this problem under control, but we do need to monitor the Varroa levels and treat as required. (More about this in Chapter 8). All of these things need to be noted on our records, and when we treat for Varroa this should be noted in the comments column of the record card, and on the treatment record card.

The last thing to check for is does the colony have enough stores to last until the next inspection. This is critical at certain times of the year. First, early in the year, when the colony is rapidly expanding it may be necessary to feed the colony. If they have sufficient stores then a tick is given in that column on the record card.

Early season feeding should be a weak sugar solution, as we want the bees to be able to use the feed without the need to collect extra water to dilute it. The type of feeder is important as well, it needs to be a contact feeder so that the bees don't have to venture far from the cluster.

During the active season feeding is not normally needed, unless the beekeeper has taken off the honey. If the weather turns wet immediately after the honey has been removed starvation can take place. I had this happen once in my early years – within a week of removing the honey, one colony was dead from starvation.

Newly hived swarms will also need feeding. They might have brought with them about 1 kg of honey, but that is soon used up in producing fresh wax for comb building during the first few days in their new home. If the weather has then turned inclement a week later they will be almost at the point of starvation. They will certainly not build up in strength and will lose the enthusiasm that is normally there in a fresh swarm.

Lastly, as we close up the hive, check there aren't any repairs that need to be done.

At the end of an apiary inspection all the individual hives will have had their records completed, but it is then necessary to make a note of anything that needs doing in the apiary next time. Do any of the colonies need another super, for example, or perhaps one colony is making swarm preparations and we need to have all the equipment to hand to do an artificial swarm? This is particularly important with an out apiary. There is nothing worse than driving a distance to an apiary only to realise that a vital piece of equipment has been left at home.

At the end of the season you can record your estimate of the honey crop from each hive. It is interesting to compare differences between hives and also between different apiaries.

Filling the record card in is one thing, but does it help us? Unless you spend a little time comparing it with the previous entries it will be of little use to you. It is just a snapshot of what is happening on the day of the inspection. Comparing it with what went before allows us to attempt to understand what is going on in the hive and what the bees might be planning.

There are other sets of records that are needed, and these are for disease control and treatment. There are requirements to maintain records of the purchase and use of treatment as stated in the Veterinary Medicines Regulations 2011 SI 2159 and these records must be kept for 5 years. I wonder how many of the 30,000 beekeepers in the UK even know of this requirement. We should also record these treatments on the record card of individual hives.

When we start out beekeeping are we taking on just a new hobby, or are we starting out on a way of life where the bees take over our lives? I had just two or three hives for many years, but having taken early retirement from the rat race, I quickly increased to 50 or 60 hives. I did this by buying bees from retiring beekeepers or from ones reducing their number of colonies and also by expansion of my existing stock. It wasn't really done with any plan in mind, it just happened, and I took advantage of opportunities when I found that bees were being sold. Similarly, with equipment, I was able to buy at very reasonable prices empty beehives that just needed a bit of cleaning up. I'll cover in Chapter 8 the pitfalls of second-hand equipment.

As a consequence of this rather random acquisition of bees and associated equipment I ended up with a very diverse quality of bees, and at that time realised that I needed to improve the stock, particularly getting rid of aggressive and/or swarmy bees.

It wasn't all bad news of course, because I did manage to produce over 1 tonne of honey in a year.

What was needed was a plan of action. At this time, I became interested in our native honeybee AMM or the British Black Bee. Many years ago the respected beekeeper and bee breeder Brother Adam had written that the British native bee had become extinct as a result of the Isle of Wight disease (I'll cover this disease in Chapter 8) which had wiped out a large proportion of the honeybee population in the UK just after the First World War. There are a small, but increasing, number of beekeepers, who now know that this is far from true. There are areas of the country that still have bees that are near native. The British Black Bee isn't really black, but it is a dark leathery brown colour. It is slightly smaller than some of its relatives but is the bee best suited to our climate. I have heard it described as being bad tempered, in fact, Brother Adam made a point of this in his book when comparing behaviour of various different races. My experience and that of others who have these bees is that this is far from being correct. They are very docile, and they work when the weather is less than ideal. They fly early in the day and continue later into the evening. I believe they also live longer than some of their continental neighbours. AMM bees are more conservative, in that they overwinter with fewer stores and have a very definite break in brood production in December and January. I wish that I had known about AMM bees earlier in my beekeeping life and had had better knowledge of genetics.

The plan that evolved was to selectively breed from my best queens, but first I had to establish whether I actually had any AMM bees. The only sure way of knowing would have been using DNA, but clearly that wasn't an option. The one tool, however, was to utilise a technique known as morphometry to study the veins in the wings (see Appendix 9). To evaluate a colony it is necessary to take the wings of 40 workers and to scan them with a high magnification scanner. Then using a specialised program the wings can be checked for certain defined characteristics that indicate which bee sub-species they belong to. The results of these tests showed that I did indeed have some colonies with a high percentage of AMM characteristics and these were used to produce queens. The whole process is very time consuming, and I made plenty of errors until I got it right. Bees' wings are small and very light, so the slightest air movements can cause problems in making up slides for scanning. Just breathing on them can mean hours of wasted time!

The use of morphometry to determine race is a well-established but lengthy procedure, but one that can be of great use. Some years ago, I collected a swarm of bees in late spring that appeared to be AMM. They were a uniform dark colour, were docile but had swarmed relatively early in the year. I let them settle in and together with the other hives in that apiary checked them using morphometry. While the rest of the colonies showed near native British bee results, the swarm results showed it to be Carniolan (*Apis mellifera carnica*) bees. I knew there were a number of colonies in the area with these queens, imported from Slovenia, often being sold to new beekeepers in the form of nucs. This Carniolan strain of bee is far more prone to swarming, which accounted for the swarm I collected early in the season. Knowing the true strain of these bees allowed me to requeen the colony with one of my own queens, to cull any drone comb, and thus maintain the quality of the bees in the apiary. It could have been a big mistake not to have tested it.

Producing queens is only half the problem solved, as they mate in the air, with drones from every hive in the area. I needed a way to ensure my new queens mated with the drones I wanted. The options to ensure that cross mating doesn't occur are either instrumental insemination or isolation mating. The first option is far beyond my abilities, and most other beekeepers as well. Living in a rural upland area of North Wales does mean that relatively close by there are areas with very few beekeepers. I found an apiary site that was located about 1200 m above sea level about 5 miles from home, in an upland valley within a forested and moorland area. With six hives with good genetic characteristics, I was able to have reasonable control over drone production. Nearby I installed up to 30 mini hives that could be populated with a small number of bees and each with a virgin queen. In this way I could produce 40 or 50 queens each year which were used to requeen my hives and also sold to other beekeepers.

In general, these queens had good AMM characteristics, those that showed an appearance to being non-native would be discarded and over a period of a few years I established some excellent, docile and productive colonies. Some years later a group of fellow beekeepers in my local association got together to help improve our local bees generally. We set up a remote mating site high on the Denbigh Moors, which I had used previously during late summer as a temporary site for production of heather honey. Unfortunately, we abandoned the location for mating after a couple of years, in favour of one closer, as it was probably far too remote for most of the queen rearers in our group. I also later discovered a beekeeper about 3 miles away who had very mixed quality of stock, so this remote site wasn't as remote as we first thought. We also helped another beekeeping association by providing larvae for their queen rearing, and the use of our mating site.

Later still, DNA testing became possible and a local university research project on many colonies in North Wales confirmed that some of my bees were indeed very near to pure AMM.

7

Queens and swarming

Selecting or purchasing queens?

Like many beekeepers I have growing concerns about the continued importation of bees and queens, leading to the introduction of disease related organisms, aggression in subsequent generations and the unsuitability to our unreliable climate of bees that evolved in much warmer and reliable conditions. Many will disagree with whether these matters are relevant, but the threat is there, so why take the risk, especially when we don't have to?

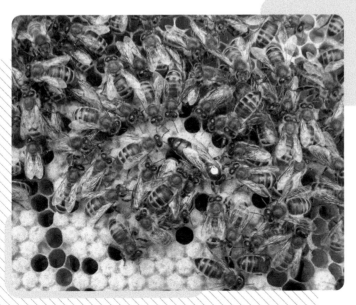

7.1
An *Apis mellifera mellifera* queen on a comb of brood.

First, let's look at why so many queens are imported into this country every year.

There are about 30,000 beekeepers in the UK and of these there are about 400 professional bee farmers, earning a living from producing honey. Some manage as many as 3000 hives. In total there are about 224,000 colonies of bees in the UK, with most hobby beekeepers having fewer than four hives.

In 2020 there were 21,405 queens imported into this country, mainly from Italy, Greece, Cyprus and Slovenia, which compares to about 10% of the number of hives. Most of these queens are purchased by commercial beekeepers. They are available early in the season and are relatively cheap when purchased in large quantities. These beekeepers can requeen their colonies and split colonies using these new queens to make up for any seasonal losses. Because the queens are young the colonies grow quickly and are less prone to swarming, so from a commercial point of view it makes good economic sense, however, it can have detrimental consequences to local hobby beekeepers.

Many of these imported queens are sold by the importers to the general beekeeping population, either as queens or made up into nucs for sale, often to unsuspecting new beekeepers.

My reason for concern is governed by the mating behaviour of queen bees, as it takes place in the open air and the queen mates with up to 15 different drones. The beekeeper has no control therefore, over the offspring of any newly mated queens.

Most beekeepers have little or no knowledge of genetics or understanding of dominant or recessive genes, so many have stumbled into buying queens and subsequently regretted their actions. If one buys replacement queens every year then you are not bothered about local mating, because the bees you have are always the same as the ones you purchased. However, the drones these colonies of foreign bees produce will be involved in the mating of other local queens, resulting in hybrid bees. 'So what?', you may ask, they will have hybrid vigour and may do very well; indeed that is often the case in the first year.

Many years ago, I saw an advert in the beekeeping press for 'Greek Buckfast Cecropia Queens'. They were described as productive and docile. I purchased a couple and they were exactly as described. I thought they were the best bee ever, so the following year I purchased four more. What a mistake this turned out to be! Three years and two generations of queens later these bees were the most aggressive bees I've ever encountered, and they would follow you for about 150 m (some six times as far as non-aggressive bees would). I know of other beekeepers who will tell similar stories of hybrid queens from different countries.

Why should that be? To answer that we need to get back to genetics. I don't want to get into an in-depth discussion on genetics, there are whole books written on the subject. To put it simply, aggression is a recessive gene, that is, it does not normally show, but if you get mating between queen and drones both with this recessive bad temper gene then one in four of the offspring will always be aggressive. It is for this reason that I would only sell a nuc headed by a 1-year-old queen, as I know that colony will not be bad tempered.

If you look at any old pictures of beekeepers they never wore much in the way of protective clothing, and certainly never gloves. Now look at how we dress. We normally wear full bee suit and gauntlet gloves, and all because of decades of imported queens.

I believe importing queens must be the greatest collective mistake in British beekeeping history and it is still going on.

We need to breed our own queens and select for characteristics that we want. We will never be able to return to a native bee population country wide, because imports will continue, but we can all try to select and improve our bees. Bees evolved to suit the climate, so it makes no sense to expect a bee from another part of Europe to do better in our climate than our own *Apis mellifera mellifera* – the native British Black bee.

Imports bring with them diseases. Almost certainly Varroa came here that way. The small hive beetle, although not here yet, has arrived in Italy probably via importation of bees. There are also many bee viruses that will be carried to the UK through importation.

It is my opinion that imports should be stopped, but unfortunately cash is king, and while we were part of the European community there were no restrictions on the movement of bees between European countries. This has changed and we are no longer part of the EU. Now importation of queens is still permitted, but colonies and package bees are not; that hasn't stopped some commercial beekeepers from trying to get around the rules by trying to import through Northern Ireland. Personally, I hope this is stopped. In fact, I would like to see imports of queens stopped as well.

Some other mistakes with queens

I killed the queen?

Not many beekeepers like to admit to this error, but I know I've accidentally killed a queen at least once. The first occasion that I know of I caught a queen between the spacers on Hoffman self-spacing frames. I found her a week later at the next inspection. The bees sorted out the problem though and raised emergency queen cells. Another time I found a queen walking on the ground almost between my feet during an association apiary meeting, which shows how important it is to hold frames above the hive when inspecting, so that if she drops off it will be back into the hive. On another occasion while leading an apiary meeting, I was just closing a hive up when someone asked, 'What's that big bee on your hat?' How the queen got there I've no idea, but it did explain why

7.2
'What's that big bee on your hat?'

we couldn't find her in the hive. The possibility of accidentally killing the queen should be a good reason for not opening a hive late in the season, or too early in the spring, as the colony has no chance to requeen. As a rule, I never open a hive after taking off the honey crop until the first inspection of the year other than to treat for Varroa. The brood frames stay undisturbed from August to March.

I lost the queen?

I've seen queens fly off immediately after being marked. The general advice in that case is to just wait for her to come back, and she will return to where she flew from. On this occasion I waited, but I didn't see her return, so closed the hive up and hoped everything would be OK. At the next inspection there she was as if nothing had happened. She was a young queen and only just mated, so she would have known her way home. Queens that are laying large quantities of eggs are too heavy to fly far anyway. There is often an assumption that because you can't see her, she isn't there.

Finding the queen is a common problem for new beekeepers and the general advice is to look for the signs of a queen not the queen herself. Many beekeepers hardly ever see the queen.

7.3
Sometimes she is difficult to see, particularly if she is busy laying an egg.

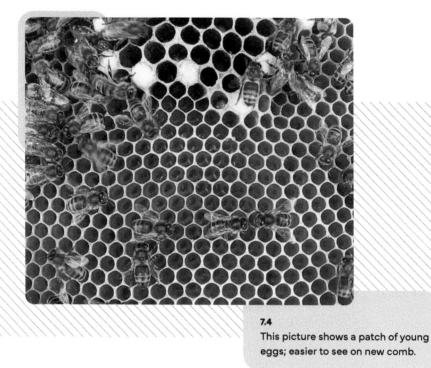

7.4
This picture shows a patch of young eggs; easier to see on new comb.

One must look for eggs, larvae and sealed brood. If there are lots of eggs she should be there, if there are few eggs then I might expect to see queen cells. If only all beekeepers could see eggs. Some beekeepers expect to see more eggs than there really are. The ratio of sealed brood to unsealed brood to eggs should be 4 : 2 : 1. At any one time, therefore, only a seventh of the brood area will be eggs.

Am I queenless?

This question normally comes when the colony has swarmed and then some weeks later there is no sign of the queen. Many times, I have been asked by relative newcomers to beekeeping if I can supply a queen, because they think they are queenless. It turns out the bees swarmed 2 weeks before and they see no sign of the queen. There would appear to be a general lack of understanding of how long after swarming it is until a new queen will start laying eggs. We need to look at the timeline. Day 1 is normally when the queen cell was sealed and the swarm appeared.

One week later the new queen emerges, and it is at least 3 days then before that queen might possibly fly. If the weather is not hot and sunny it could be up to 3 weeks until mating takes place and then a further 2 or 3 days until she lays the first egg. This indicates that 4 weeks is a good guideline before you need to even look in the hive. My general advice is, don't open the hive for 3 weeks after you think the queen has emerged. Keep an eye on what is going on at the entrance. If you see pollen being collected in quantity, then is the time to look inside, because they will only be collecting pollen if they have larvae to feed.

The only way to be sure that the colony is queenless is to insert a 'test frame', that is, a frame of brood containing eggs and very young larvae from another hive. If the colony is queenless the bees will start emergency queen cells on that comb.

Sometimes a queen becomes a 'drone layer'. Our British summer weather is not always conducive for queen mating. If the queen has been restricted from flying by inclement weather, eventually she starts to lay but is only able to lay unfertilised eggs. If the beekeeper can notice this then it might be possible for him/her to rectify the situation by introducing a new queen. Unfortunately, by this time all the workers will be old and often the colony isn't worth trying to save. It might be best to remove this queen and unite the hive to another that has a laying queen.

'Laying workers' is a situation that even experienced beekeepers find difficult to diagnose. Generally, this occurs in one of two scenarios. The first occurs at the beginning of the season after the colony comes through the winter and for some reason the queen has died. The bees have no means of raising a new one, as there are no fertilised eggs. As there is neither queen, nor brood pheromones present some workers have the ability to lay eggs. This egg production would normally have been suppressed by the presence of pheromones. Unfortunately, all the eggs laid by these workers will become drones.

The second situation when this phenomenon is seen is during the season, normally after swarming has taken place. The new queen might become lost or due to some intervention by the beekeeper there is no queen. The bees don't have the ability to produce a queen under emergency conditions as there are no eggs. The colony becomes hopelessly queenless and laying workers start to lay eggs resulting in drones. At this point the colony is probably small, with only old workers, and almost impossible to requeen.

One sign of laying workers is multiple eggs in cells, and I have seen this with a very young, newly mated queen. The new queen may sometimes start laying drone eggs. Both of these things seem to settle down after a few days.

Marking queens

Queens that are marked are much easier to spot in the hive. Many beekeepers use the five international colours to signify the year of the queen.

COLOUR	YEAR ENDING
White	1 or 6
Yellow	2 or 7
Red	3 or 8
Green	4 or 9
Blue	5 or 0

I remember the colours W,Y,R,G,B as – Which Year Raised in Great Britain, others use –Will You Raise Good Bees.

Some beekeepers only use white and I know of some who alternate between white and yellow. The problem with yellow is that many returning foragers have yellow pollen loads, and you see yellow dots everywhere when looking for the queen with that colour spot.

It is very easy to get marking wrong. I like to use a water-based marker pen, readily available from the beekeeping suppliers. Apply too much and the queen ends up covered in paint, too little and the paint soon wears off. The trick is to put just enough on and to ensure it is painted actually onto the back of the thorax of the queen and not just onto the hairs. I always test the pen on another surface first before marking, to make sure there isn't too much paint on the end of the pen. In hot weather or even just in your pocket, the pen can become warm and the contents pressurised. It you are not careful, when used it could spread paint everywhere. Red, green and blue are not as easy to see, particularly blue, so I can see the reluctance by some beekeepers to using these dark colours.

Our village has an annual summer show, and for many years I had a stall selling honey and other produce. I normally had an observation hive, which always proved popular. I recall one lady who asked which bee was the queen. I replied, 'She is the one with a red spot on her back'. Almost at once she was able to find the bee with the red spot. The following year this same lady came to the stall and said, 'I know which one is the queen, because she has a red spot!' I explained that red was last year's colour and this queen had a green spot. She was amazed and said, 'That's wonderful, but how do the bees do that?' Once I explained about who did the marking, she soon found the queen, resplendent with her bright green spot that I had applied the day before.

Some makes of marking cage are better than others. There are some made entirely from plastic, and with these, it is impossible to see what is going on under the mesh. The major disadvantage to this type of cage is that it is only too easy to stab a bee with the sharp needles. For this reason, I often jokingly refer to it as 'the stab the queen cage'. Used carefully it is very effective in holding the queen immobile to be marked.

7.5
There are various cages available to hold the queen while marking, and my favourite is the Baldock cage or crown of thorns as some call it.

It can cause damage to brood comb, so if possible, the queen should be trapped on an area of comb away from the brood. You can pick up the queen with your fingers to mark her, but this is not a job for the faint hearted, as you can only do it with bare hands. It takes a lot of skill to catch a queen and pick her up from the face of a comb covered in bees! Get it wrong and you will get stung by the workers or you may damage the queen. The queen will not sting though; it is almost as if she is pro-grammed not to do so. She will only ever sting another queen.

There are other methods of marking, including stick-on numbers. I've even seen metal stick-on discs that allow the beekeeper to pick up the queen with a magnet.

Decades ago, I saw typewriter correction fluid 'Tip-Ex' being used for queen marking. It was certainly permanent, but the solvent used was toxic and strong smelling. A marker paint should be as low odour as possible to avoid the bees killing the queen once you have marked her.

Clipping the queen

This is another job that requires practice, and you might wonder why some beekeepers do it? For a couple of years, I clipped a wing on all my queens, but only because it is a required procedure for the BBKA General Husbandry Assessment. The advantage of clipping a queen is that if the bees swarm the queen can't fly properly and will fall to the ground. The rest of the swarm know that she is not with them and return to the hive. The colony will then swarm with the first virgin queen to emerge. The supposed advantage to the beekeeper is that the time between inspections is increased from 1 week to 10 days. The disadvantage that I find is that as they had swarmed and lost the queen it is then impossible to do an artificial swarm, so to me swarm control actually became more difficult. I don't clip queens anymore, and I have a philosophy now that if I do lose a swarm then one of my locally bred, good tempered and near native queens is somewhere local to me producing drones for my queens to mate with, so all is not lost. The other potential problem with clipping queens is that unless you manage to find the clipped queen from the swarm on the ground near to the hive, that genetic line may disappear altogether if the new queen is lost or doesn't get mated.

Clipping the wing requires practice! When holding a queen prior to clipping her wing we need to be patient. When the queen feels the scissors on her wing she will often raise a hind leg to brush them away. If you don't wait until the leg goes back down then you will clip a leg as well as a wing. One needs to remove a third of the wing and no more. General advice on practising clipping is to clip drones, as you can't practice with workers because they will sting. Drones don't sting, but when you hold them as you would a queen they don't keep their wings still – they buzz!

It is possible to clip a wing using the marking cage mentioned above. Having marked the queen if you gently lift the cage edge that is nearest the queen's tail, she will back up, and if you are lucky a wing will protrude through the mesh. At this point you can clip the wing and if you are even luckier you won't have snipped through the thread of the cage. (I also became adept at repairing the cage!)

Considering all these factors, I have concluded that for me it is a mistake to clip the wings of queens. I have wondered why it is a requirement of the BBKA to carry out this procedure if one wishes to become a master beekeeper?

Prevention of swarming – the impossible dream

Swarming is the natural way that honeybees increase their numbers. As responsible beekeepers we should try to avoid letting our bees swarm and this is one of the most difficult lessons a new beekeeper needs to learn. Even those of us with years of experience get it wrong. We will never understand honeybees if we try to apply human logic.

We need to try to understand what triggers swarming if we are to stand any chance of preventing it. There were many theories to swarming in the past, but it is now generally accepted that the triggers for it are overcrowding and a reduction in levels of queen pheromone in the colony. The amount of queen pheromone (sometimes referred to as queen substance) produced by the queen reduces year by year, and the bees spread what pheromone there is around the hive. The colony grows in size as the season progresses then the dose of pheromone per bee diminishes and the balance in the hive changes. This in turn triggers the bees to build queen cells. What makes the queen lay in those cells is unknown, but it is probably some sort of feedback from the worker bees. There has been suggestion in the past that workers transfer eggs into queen cells, but I'm not sure that there is any proof of that. There are various things we can do to reduce the tendency to swarm, which we might term swarm control. These measures rely on maximising the queen pheromone dose per bee. If we only keep very young queens it will mean that the production of queen pheromone is at its highest (this is the philosophy behind requeening early in the season). We also need to ensure that the colony does not become overcrowded. Beekeepers therefore need to ensure the brood box is of sufficient capacity for their particular strain of bee, and also to provide sufficient room in the hive for honey storage. Swarm prevention is trying to ensure that a swarm does not occur, and taking action once swarming preparations are underway.

It is worth noting that once swarm preparations have been started they are almost impossible to stop. The beekeeper just has to work with the bees to ensure a swarm is not lost and a new queen is produced. For a swarm to occur there are three things that need to be in place. There has to be a queen (not necessarily a mature queen, it could be a virgin queen), queen cells and sufficient number of bees. Additionally, the weather has to be warm enough, so swarms will often come out of a number of hives that are in different stages of swarm preparation if the weather suddenly turns hot and sunny. Swarm prevention methods are designed around regular inspections and when action is needed to then separate the queen from the queen cells.

During the swarming season we need to regularly inspect our colonies, normally on a weekly basis to check if swarming preparations are taking place.

You may ask, if this is a natural thing for bees to be doing then why should we try to stop them from swarming? There are many reasons why we don't want a colony to swarm. First, it can be a nuisance. Many people are frightened by a swarm; indeed, it can be quite awesome. I've heard of two different accounts of children being rushed in from the school playing fields because a swarm has passed over. Swarms can end up in places that are at best inconvenient and at worst a nuisance to the general public. Most summers there are accounts in the media of swarms in unusual places 'terrorising the public'. From the beekeeper's point of view, swarming can have a potentially devastating effect. The bees, when they swarm, first gorge themselves on honey so that they have enough food to last until they set up in their new home and build new comb. Workers in a swarm carry on average almost four times the amount of honey in their honey stomachs than normal, and I guess that this probably amounts to almost 1 kg of honey. Taking the value of honey to be £10/kg this amounts to a loss of £10 of income to the beekeeper. In a very poor year this might represent a significant part of all the season's honey crop. The swarmed colony will have lost a significant proportion of its bees, there will be a delay in the new queen starting to lay eggs and they will seldom produce any further surplus honey in that year.

I've heard it said that during a lecture only about 10% of the information is actually fully taken in by students. I shouldn't have been surprised when I once had a phone call in November from a student on that year's spring beginner's course, to say they thought they were queenless – what should they do. My first question to them was, 'What makes you think that?' They replied that they had that day looked through the hive and could find no sign of brood or the queen. Wondering why they should have thought it necessary to mess about with a hive in November, however good the weather, they answered, 'You told us to regularly check for queen cells.'

Was that their mistake, or mine for not explaining it correctly?

It is clear to me that beginners find it difficult to see eggs, can't find the queen and don't always recognise queen cells, so it should be no surprise that they don't always pick up the signs of swarm preparation.

To make things simpler for beginners I have a plan that should make things easier.

1. Look for drones (everyone can see drones). Once you start seeing drones in the spring then is the time to start weekly checks for swarming. Why? Because the bees don't start to think about swarming until there are drones about. From a timing point of

view a drone takes 24 days from egg to emergence, and then it is mature enough to mate after about three weeks. The queen, however, takes only about 8 or 9 days from egg laying until swarming, and a new queen will emerge 1 week later and be ready to mate 3 or 4 days after that. The bees have to plan in advance, so no drones, no swarms.

2. Don't spend your time looking through the hive trying to find the queen. They run about, hide under other bees and are generally difficult to see, even if they are marked (and you have remembered to put on your reading glasses).

3. Look for queen cells. They don't move and are relatively easy to spot once you know what you are looking for.

7.6
The cell at the bottom of this picture is a fully developed and sealed queen cell. A further unsealed queen cell is on the right-hand side towards the top. The other large sized cells on the bottom right corner of the frame are drone cells. You will need to move bees from the edges and corners of the frame to see cells.

4. Once you see a queen cell, if it is not already sealed, look inside. Is it empty? If so, no need to worry. Does it have an egg in it? Answer – 'Don't know – I can't see eggs!' Should have been to Specsavers perhaps! Still no need to panic, check through and see if there are any more queen cells. Only one? Perhaps a super-cedure cell – you might cut it off, at least then you know you are safe that they are not going to swarm (provided you have seen the queen). Does it contain a larva?

If this is what you see, then, yes you need to take action now! Check how many other queen cells there are and decide what course of action is required. **Whatever you do, don't cut out any queen cells at this stage.** If the queen cell is sealed, then it may be too late. The colony may already have swarmed.

7.7
This queen cell contains a small larva which is probably two days old. The cell will be complete and sealed in about 3 days' time.

The normal recommendation would be to do an artificial swarm which always starts with the requirement to 'find the queen', but we already know that the beginner, and often the more experienced bee-keeper can't find the queen. At this stage there appear to be three options that beekeepers might adopt.

1. Let them swarm.
2. Cut out all the queen cells.
3. Somehow do your proper swarm control irrespective of whether you can find the queen or not.

Option 1. This shouldn't really be a serious consideration, but sadly all too often it is!

Option 2. This will only delay the swarm as the bees just build even more queen cells, and if you miss just one they will swarm anyway, within just a few days.

Option 3. This is the only real option, but why do so many of us get it wrong? (Some of us do more often than others.)

The most widely used method of swarm prevention is to do an artificial swarm. This is not the only method of dealing with a colony making swarming preparations; in fact, whole books have been written on swarming. Most methods rely on finding the queen and that is probably why so many beekeepers fail to carry out a proper swarm prevention.

I've never seen this variation in any book, but I use a foolproof way of dealing with the situation without needing to find the queen. See Appendix 1 – Artificial swarm without finding the queen. It is possible to use the first part of this method to isolate the queen and then just make up a nuc with the queen, if that is your preferred method of dealing with swarming.

I know of beekeepers who split their hives, by taking off a queenright nuc, prior to the bees starting queen cell production. The colonies make emergency queen cells and apparently there is no need to reduce the number of queen cells when they have been raised under emergency conditions. A word of warning though, if the colony has already started making queen cells then they are already in the swarm preparation state. In this case you will need to make an inspection 1 week later and remove the excess queen cells.

What happens if even with the best endeavours of the beekeeper the bees still manage to swarm? At this stage it is damage limitation. We have already made the first mistake of letting them swarm, what we do not want to do is mistake number two – let them do it again a week later!

Some years ago, I had a number of phone calls all along these similar lines. Question from caller: 'Do you have a queen?' Rather than just offering to sell them one I would ask why? Answer: 'I was on holiday/ too busy to look at the bees and they must have swarmed. So when I inspected them I cut out all the queen cells and now I think they are queenless!' Alternative answer: 'When I went through the colony 3 weeks ago there were lots of queen cells and I cut them out. Now I can't find the queen!' When asked if they had swarmed, the reply was often, 'I don't know'. In these situations the beekeeper may have effectively killed the colony. There seems to be a widespread and perhaps perverse obsession with cutting out queen cells. Some beekeepers will do it as a matter of course. At worse you end up queenless or with a poorly produced queen that the bees have managed to make from an emergency cell.

Once in my role as a bee inspector while inspecting a colony I saw a queen cell on a comb. This of course was a very common occurrence during the summer, but on this occasion the beekeeper reached across with his hive tool and cut off the cell saying, 'I don't think we want that.' I asked what would be the result if that were the only queen cell in a colony which had lost its queen. He hadn't thought of that. On this particular occasion there was no sign of the queen, or eggs, but there were other queen cells. The colony looked to me as if it had swarmed within the past 2 days, judging by the number, or lack of, eggs in the hive.

If you know a colony has swarmed, then go through the hive and check how many queen cells there are before you touch any of them. Then leave just one cell, preferably an unsealed one, and remove all the others! It is easy to damage a cell when removing the frames, so I try to leave a cell that is located near the centre of the frame if at all possible. Some beekeepers are wary of leaving just one queen cell and prefer to leave two. My advice is if you must do that only ever leave unsealed cells and ensure they are the same size (that is, the same age). I once heard of a teacher of a beginner's course advocate leaving two queen cells – one big one (sealed) and one small one (unsealed). These would be different ages, so when the first queen emerged the colony would have both a queen and queen cells, which are the prerequisites for a swarm, and they will produce a second, or cast, swarm. What I learnt from this is that even some apparently experienced beekeepers don't properly understand their bees, or the principles of swarming.

You may come across a colony that has swarmed and only has sealed queen cells. My advice is to leave just one that looks the best and is somewhere near the middle of a comb, rather than along the bottom edge. If the colony swarmed a week previously you will find one or more opened and empty queen cells. You might see queens emerging as you look. If you find emerging queens, carefully open all the remaining queen cells and let the virgin queens loose into the hive. The bees often hold emerging queens in their cells until after a cast has left. Don't leave any cells of any description. If you don't do this they will swarm again, and again until there are no queen cells left. If you release all the virgins, they will sort themselves out. Just leave them to it and have another look in about 3 weeks' time.

When a queen is about to emerge from the cell the bees thin off the wax around the end, to make it easier for the queen to come out. The queen cuts a neat flap at the end of the cell to emerge and at this stage a worker bee may go into the cell to clean it out. The other bees sometimes push the flap back and re-seal it. I've often seen sealed queen cells in a hive long after the virgin queens should all have hatched. If you break these off there is often a dead worker bee facing towards the top. I once heard a beekeeper discussing problems with queen rearing and suggesting that he was finding dead, stunted queens facing the wrong way in queen cells. Isn't it strange how sometimes we fail to see the obvious.

You may be lucky to see your bees swarm, but generally they wait until you are not there. They appear to be more likely to hang up in a cluster in your neighbour's garden, hardly ever in your own garden. They wait until you are about to collect them and then just disappear over the horizon never to be seen again. It would seem that the ownership of a swarm remains with the beekeeper, provided he can keep it in sight. Traditionally the swarm would have been attracted by banging on a saucepan, which is called tanging. Whether this worked I have no idea, but at least neighbours would know that there was a swarm about. From bitter experience I can only guess that in days of old, beekeepers could run fast because a swarm flies at about 7 mph. It's no wonder I could never keep up with mine when I tried. Assuming your luck is in and you can collect the swarm, what to do now?

General advice has always been to check which of your colonies has swarmed and then house the swarm on the site of the hive they have come from, moving the old hive about 3 ft (1 m) to one side. This will result in more bees with the swarm and the old hive being depleted as the foraging bees return to the old site and join the swarm. This system is supposed to maximise the honey crop. When I've made this suggestion to beekeepers in the past, often the response is, 'How do I know which one has swarmed?' Well you certainly aren't going to know just by looking at the outside, unless you actually saw them come out. You need to open all the hives in the apiary and check for queen cells. I did this once on an out apiary when I found a swarm hanging on a nearby tree. None of the hives had swarmed! The swarm must have come from elsewhere – I later discovered another beekeeper with about 12 hives less than 400 m away, no wonder my honey yields at that apiary had gone down, but that is another story.

"From bitter experience I can only guess that in days of old, beekeepers could run fast because a swarm flies at about 7 mph."

Collecting swarms

Collecting swarms is a whole new experience for a beekeeper, and I have collected a great many. Don't forget that most people know nothing about insects, except that bees and wasps sting. If you want to put your name on a list to collect swarms be prepared for phone calls to get rid of wasp nests, bumble bees, solitary bees and only very occasionally swarms.

Difficult to keep one's composure having driven 8 miles (13 km) to find the 'swarm' is in fact a cotoneaster bush, in flower, covered in foraging bees, all of them content to get on with life and they aren't bothered what else is going on in the garden.

7.8
A large swarm on my pear tree.

A few years ago, I had a phone call about a 'swarm' of bees on a shrub near the front door of a house. Wise now that this might be another cotoneaster story, I asked a few questions. Sure enough it was a shrub of unknown type, small red flowers and covered in 'literally' thousands of bees. 'No' there were not any bees there at night, but they 'were back again the next day. Something has to be done because the postman is refusing to deliver'. I offered my advice, which was to wait until dusk and then cut down the bush and dispose of it. 'Can't do that it's a lovely shrub', was the reply. My answer, 'OK then, put up with the bees and talk to the postman'. That's the first time I had heard a postman story that was bees preventing delivery, it's normally dogs!

So often people have contacted the council about nuisance bees and have been recommended to contact a beekeeper. The mistake we make as beekeepers is to try to help, as if we are a free pest control company. There is a general assumption that beekeepers will collect swarms of bees, get rid of colonies of wild bees or wasps all free of charge because we are beekeepers and that's what beekeepers do! Do plumbers and electricians do emergency call-outs free of charge?

Of course, we should only try to collect a swarm if it is safe and easy to do so. Now that I'm getting older the days of climbing ladders or trees to recover someone else's bees are gone, although perhaps foolishly I've done this many times in the past. Better probably to just say, 'Sorry my public liability insurance doesn't cover for working at height', that is received better than the blunter alternative, 'I'm a beekeeper not a steeple jack!'

I once had phone calls on two successive days regarding bees in a chimney. The first one I wasn't surprised to receive, as earlier that day I had been alerted to a swarm hanging from a branch of a tree above the path leading to the entrance to a dentist's surgery. I wondered at the time whether I should shake them into a box there and then or leave them until business was closed. The first option could have caused problems to patients at the dentist and to passers-by, so I decided to leave them until the evening. Just after lunch I received a call from the town hall, about 200 m from the site of the swarm, to say they had bees down the chimney. On my way to the town hall I called in to look at the swarm I had seen that morning. They had gone, and I guessed that they had moved into the chimney of the town hall.

The situation there was that bees were going down the chimney and coming out of a vent at the bottom into the room where they collected rents. They had had to close the office. I took off the cover to the vent and could hear the bees up the chimney. I thought that as they hadn't been there long perhaps a bit of smoke would get them out! (I'm sure that I had read that somewhere!) I lit my smoker and puffed smoke up the chimney, hoping it would drive them out. The buzz of the bees increased but then the smoke came back down the chimney. At this point the fire alarm went off and with great excitement the building had to be evacuated. I had successfully cleared the building of people, but the bees were still in the chimney and still coming out into the room. Clearly it was time to put Plan B into operation. The bees liked their new home, which the scout bees had taken time and effort to select while the swarm was hanging around all morning near the dentist's surgery and a bit of smoke wasn't going to get them out. I reluctantly came to the conclusion that on this occasion the bees had won! The only thing left to do was to stop bees falling down the chimney and coming into the room, so I stuffed lots of newspaper up the chimney, and borrowed the vacuum cleaner from the caretaker to suck up the bees that were now trapped in the secondary double glazing.

The following day was bees in chimney 'Take 2'. A swarm had just gone into the chimney of a large country house and bees were coming down into a bedroom and gathering on the window. We drove down the long drive with trepidation following the previous day's fiasco (I was accompanied by my wife this time to stop me doing anything silly). On arrival we were met by the handyman who led us through the mansion past the suit of armour at the bottom of the stairs, to the master bed-room. Plan A was to try the smoke trick again, surely it would work this time! However, it had the same result, but no smoke alarm this time. There were sooty bees falling down the chimney, walking across the cream coloured silk rug and then flying up to the window and leaving black, sooty marks on the white window blind. Clearly this needed an alternative approach, but I couldn't think of anything. The only thing to try was just to stop these sooty bees coming down the chimney and causing even more damage to the valuable furnishings. Fortunately, there was an adequate supply of the *Financial Times* newspaper, provided by the handyman, so I stuffed the chimney to stop the bees falling down.

That had stopped the bees coming into the room but looking across to the window through the fog of smoke from my smoker I could see that there was quite a large cluster gathering on the glass. I went back to my pickup to collect a nuc box, and get some fresh air into my smoke filled lungs.

I collected as many bees as possible from the window into the nuc box and left it on the windowsill until the evening. Result of all this work was a blackened carpet and window blind, a room that stank of smoke, a small queenless nuc of bees, the main swarm still in the chimney and two bee suits which had gone from white to patchy black and needed to go straight into the wash again!

The owner was very gracious about the whole thing and pleasantly said not to worry as the room needed redecorating anyway! Some years later he decided to take up beekeeping himself and called me to ask for advice.

I might add, somewhat guiltily, that I did get paid my expenses for both these failed attempts, but the lesson is that if there are bees in a chimney then that is not the job for a bungling amateur beekeeper. Future calls where bees had gone into a chimney are now responded by suggesting that the owner blocks it up, to stop the bees coming down. If the chimney is one that is being used then sadly they will have to call in a pest control company with a cherry picker, because ... 'Sorry our public liability insurance doesn't cover for working at height'. This really is a great excuse!

7.9
This swarm was at the college where our local association has their apiary. Obvious where it came from then!

This swarm on a lamppost required the use of a ladder (working at height again) and someone to hold the ladder. The college caretaker volunteered and was suitably attired in my spare bee suit. This is a must, as if you accidentally drop a clump of bees down on your assistant they instantly let go of the ladder if not suitably attired!

Another favourite place for a swarm to take up residence is a compost bin, but mostly it is bumble bees that take up residence here rather than honeybees. I did once manage to rescue honeybees in late September that had been there since July. I then put the whole assembly into a hive, with more already drawn-out frames.

7.10–7.11
Bees in a compost bin.

7.12–7.13
I cut out the combs and fitted them into frames and tied the combs in with garden string.

The hive was placed on top of the compost bin and a feeder put on. It was collected a couple of weeks later. At the next inspection, intending to remove the string, the bees had all settled in, but the string that held the comb in place had disappeared! I can only assume that the bees didn't like it in the hive so chewed it up and threw it out, although I never found any evidence of it. They survived the winter and were transferred onto new combs in the spring.

Note to self – next time use wire.

Other interesting *rescue* jobs

I was asked if I could remove some bees from the roof of a building in a nearby village. The owner was going to renovate the house, and was allergic to bees but he didn't want to harm them. I agreed to get them out if he would provide safe access with a scaffold tower.

7.14
Working at height on a scaffold tower.

7.15
The colony had established themselves in the space under the soffit of the roof and had built quite substantial combs.

The soffit was carefully removed, and the combs cut out. Those combs that contained brood were trimmed to fit into empty frames, fixed in place with wire, and then put into a nuc box. As many of the bees as possible were shaken into the box which was then left on top of the roof for the bees to settle in.

7.16
Putting the frames into a nuc box.

7.17
Putting the nuc box on the roof.

I think half the bees were on my head and back by then. The next evening when all activity had finished they were moved away. All successfully done, although I probably shouldn't have been working at height. The bees were put into a full hive and fed. A week later wires holding the combs were removed. Finally, the following year they were transferred onto new frames.

Another interesting rescue was late in September when I received a phone call about a swarm of bees in a tree near one of my out apiaries. I couldn't believe it was a swarm so late in the season.

On investigation it was found not to be just a swarm, but had six combs attached to the underside of a branch. Carefully each comb was cut off the branch, with as little disturbance as possible, and then fitted into frames. This time I used wire to keep the combs in place.

The frames were put into a plastic hive together with five drawn-out combs with stores. While doing this I saw the queen, which had been marked. The hive was then strapped on top of the branch, just above where the cluster had been. The bees could then fly in and out, close to their original cluster location.

7.18
'Swarm' in a tree.

7.19 Working at height again.

"On investigation it was found not to be just a swarm, but had six combs attached to the underside of a branch."

7.20 The combs with brood were carefully fitted into frames using wire.

7.21 The hive was strapped on top of the branch.

The hive was left there for 2 days before being brought back to my home apiary in the evening. There was brood and the queen had been marked, and I pondered where these bees had originated and guess that they had come from my apiary some 25 m away. I had lost a swarm there during July and the presence of a marked queen of the same colour seemed to confirm that. My mistake was allowing them to swarm, but a bit of a bonus to get them back 3 months later.

A bit of advice about putting a collected swarm of bees into a hive, never do it in the dark. This might sound obvious, but it certainly wasn't to me. If you collect the swarm properly and leave it until all the bees have settled before taking it away, it may be quite late before you get to the apiary. Bees don't fly in the dark, but they do walk and if you tip out your swarm into the hive you are sure to get bees on you which will stay there. As you get back into the car you realise you have brought a few dozen bees with you. It is much better to leave the swarm in your collection box until the morning and sort everything out in the light.

If you want to hive the swarm in the traditional way, by running them into the entrance to the hive, remember that the ramp must incline up to the entrance, as bees walk up the slope. No slope and they will stay where they are. The second thing to remember is that they will walk from the light up into the dark interior of the hive, so it doesn't work too well if it's too late in the day and has gone dark. If you do, they may still be clustered outside in the morning.

"If you want to hive the swarm in the traditional way, by running them into the entrance to the hive, remember that the ramp must incline up to the entrance, as bees walk up the slope."

It is probably worth watching while all the bees go in, as sometimes they can come straight back out again. My preferred way of hiving a swarm is to take the middle four or five frames out of the brood box and tip the bees into the gap. The remaining frames are then put back in, but do not push the frames down, wait until the bees start to climb up and the frames will gently settle down. Then close up the box. This method also has the advantage that if you want to ensure that the bees stay, then you can put a queen excluder under the brood box. This is particularly useful when dealing with cast swarms which often abscond from the hive immediately after being hived. Leave the queen excluder in place for a couple of days before removing it, but don't leave it there for any length of time because if there is a virgin queen she will need to get mated very shortly. I once thought it might be easier to put an empty super on top of the brood box and just tip the bees in. When I looked the next day, the bees were drawing out wild comb in the super rather than in the frames of the brood box. They obviously didn't like a void above the frames. I've also had them draw out wild comb in the roof space above the crown board, so always best to close the feed hole as well.

It can be a bit annoying if repeatedly you get calls about swarms all within a few hundred yards of a particular beekeeper. I've even heard of one swarm collector who called at the same address 3 years in a row, before learning that it was a beekeeper who lived there who couldn't be bothered to collect his swarms so called someone else out to deal with them!

I have had to deal with a swarm on a tree in the high street, only about 100 m away, as the bee flies, from a local beekeeper. Interestingly I had quite an audience, across the street, as I shook the bees into my box, and then waited by the box until they had all settled in. I find it best in these situations to not bother with gloves and to remove my veil once they are in the box. That way, by showing that they present no problem to you, there is less chance of panic with people in the street. After a while I don't think passers-by realised that I was standing next to a box full of bees with a few stragglers still settling in. I had also provided entertainment for the customers of the cafe just across the road.

In spite of following all the advice on weekly inspections, bees can and will swarm, and artificial swarms will not always work out as planned. The weather in the early part of 2021 caused me unfathomable problems that I had never encountered before. The winter seemed to drag on and then in April we had a long period of dry, but cold weather. The bees built up slowly but didn't bring in the normal quantity of honey. This was followed by a cold wet May, and some of the colonies even needed feeding; something I had never needed to do previously at this time of year. At the end of May the weather changed and daytime temperatures rose from around 10°C to 20°C or more. The bees suddenly decided it was time to swarm.

An inspection of one hive showed that they were making queen cells, and there were a number of queen cells containing larvae. I did a standard artificial swarm with the queen and one frame of brood (with no queen cells) plus new frames with foundation, on the site of the old hive. The other part, with the rest of the brood and one remaining queen cell was moved to one side. Two days later I had a swarm, but wasn't sure which hive in the apiary they had come from. The swarm hung up for only about 30 minutes before disappearing. Checking through the hives all the queens were present apart from in the artificial swarm, which now contained a small queen cell. Perhaps, had I known that this might happen I could have put a queen excluder under the brood box of the artificial swarm, but I had never thought that it would have been necessary before.

An inspection 5 days later on a Monday showed that none of the other colonies were making swarm preparations. One had queen cups but no eggs, one had queen cups with eggs but no larvae. As a precaution I carefully checked all the frames and removed all the small queen cells and cups. On Wednesday two of the colonies swarmed! I was in the garden so was able to see which hives the bees had come from.

The first swarm conveniently landed in my pear tree and the second one 30 minutes later landed in the cherry tree. (These bees had obviously not followed the advice of their predecessors which had flown past my trees into my neighbour's garden.) Swarm number one was dealt with by placing a nuc box with just one frame containing honey above the cluster, and then waiting for them to climb up into the box, during which time I saw the queen.

They were all in the box by the evening and were put into a new hive on the site they had come from, with the brood box full of frames with foundation. The bees were fed to encourage them to draw out the comb.

Swarm number two was hanging in an elongated cluster, which I shook into a nuc box. They immediately came back out and rather than going back to where they had been clustered, flew into the air and re-gathered about 25 m away on my neighbour's hedge, next to the road. All the neighbours disappeared indoors, and all their windows were shut, in spite of the hot weather. I collected them by placing a nuc box with a frame containing some honey above the cluster and by the evening they had all gone into the box. They too were hived on the site they had come from with foundation and fed.

The next day, Thursday, swarm number two came back out and clustered in my hedge. They were collected by shaking into a box and in late afternoon were put back into the hive with a queen excluder under the brood box.

That same day I inspected the hives that the swarms had come from and although there were queen cells they only contained eggs. There were some queen cells that looked like emergency cells, that is converted worker cells, that had very young larvae.

The following day, Friday, swarm number one came out again and hung up in the pear tree, exactly as they had done two days earlier. I tried to attract them into a box, but within 30 minutes they took off and disappeared over the horizon. Checking on the hive they had come from, most of the foundation had been partially drawn out and there was a patch of eggs on both sides of one frame. A relatively small number of bees, just a few thousand, remained. An inspection a week later revealed about 12 queen cells, which were cut down to just one unsealed cell.

What about swarm number two? They too tried to swarm on Friday, but as the queen was contained in the hive by the queen excluder, they gathered on a bush and then drifted back into the hive. An inspection 7 days later revealed that most of the foundation had been partially drawn out, and there were eggs and unsealed larvae on five frames.

7.22
Nuc box fixed above the swarm when located in a difficult position.

During this same period the other hives in the apiary were all making queen cells and swarm prevention measures were needed. In these hives I separated off queenright nucs and left the colonies to make queen cells which were then reduced to only one in each hive 1 week later. It is vital that you don't miss this follow up inspection and removal of excess queen cells if you want to avoid secondary swarms.

About 2 weeks later we heard the noise of a swarm again, and sure enough lots of bees took to the air and started to cluster on one of the fruit trees. Then about 15 minutes later they started to drift back to the hive. I can only assume that this was a 'mating swarm'. An inspection 1 week later showed that there was a laying queen in residence.

What can we learn from these strange happenings? First, the advice on swarming and swarm control will work most of the time, but not always. Second, previously only my immediate neighbours knew I had bees in the garden, now the whole street knows, they thought I had given up beekeeping a couple of years ago when I had a big sale of equipment. Third, all beekeepers will have tales of swarms that have not followed the books. Jokingly I've often heard beekeepers say, 'My bees obviously haven't read the books.' All I can suggest is that while the books are not necessarily wrong, they fail to say that it doesn't work every time. Looking back to those times when I had 60 hives in about 10 out apiaries, inspections tended to be less frequent and I know that some hives swarmed. I wonder now how many artificial swarms that I did actually worked fully. My preferred method of swarm control then was to take off a nuc with the queen and allow the colony to make a new queen (going back into the hive after 7 days to reduce all the queen cells to one). This generally worked well, and required far fewer bits of equipment. I always carried a few nuc boxes with frames with me on my visits to out apiaries.

Perhaps some of the problems with artificial swarms might be explained by how they differ from actual swarms. The make-up of a swarm consists of the queen, with about one-half the worker bee population of the hive. The artificial swarm is much the same, but the age of the workers differs. I've heard it taught that the bees in a swarm are the older flying bees, and that is wrong. The workers in the swarm are younger house bees, particularly bees at the right age to be making wax and drawing out comb. Compare this to the age make-up of the artificial swarm. Here we have the old queen, a small amount of brood with a small number of young bees on the comb and ultimately nearly all the old foraging bees.

Is it any wonder then that the artificial swarm sometimes fails? Similarly, if the swarm is hived on the old site and then is joined by the foraging bees the balance is disturbed, and they still have much of the swarming incentive remaining, as they are still in the same place and still have the foraging bees. I don't have the answers, but what I do know is that when it all goes wrong and the bees swarm, sending the neighbours running for cover from their garden, your credibility as a beekeeper is stretched to the limit (as is the neighbours' patience). I suppose the mistake is just to do things according to the book and assume that everything will be fine. If only life and bees were that simple!

8

Pests and diseases

There are numerous things that can affect the health and well-being of our honeybees, some can be seen with the naked eye, others only under a microscope and viruses that are too small to be seen even under a high-powered microscope.

Fortunately, in the UK we don't have a problem with large predators like skunks or bears, which can be a nuisance in North America.

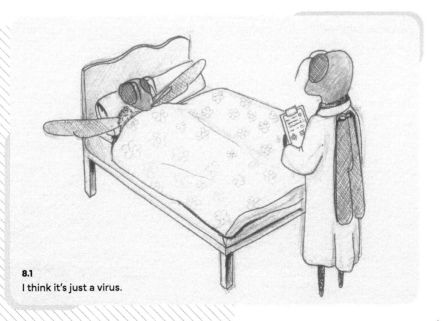

8.1
I think it's just a virus.

In normal circumstances, the largest animal which might cause a problem is the badger. The only occasion that the badger has caused any significant damage to my bees was to a five-frame nuc that was torn apart overnight, and all the brood and stores eaten. I've seen evidence of badger activity around hives in an out apiary, with scratch marks near the hive entrance. I think that a full hive is just too big and stable for a badger to knock over. A bigger problem is actually farm animals.

I once had an out apiary in the corner of a field on a nearby farm. The field was used for arable crops, so had no stock grazing. The farmer and I agreed that it wouldn't be necessary to fence off the site and all went well for a couple of years. I visited the site in January to check the hives for stores, only to find one hive knocked over and trampled on. Checking with the farmer, he told me that a neighbouring farmer had transferred some sheep from a field on one side of his farm to the other; they had in fact made the journey over three days, grazing on the way. Difficult to say which mistake was the biggest, first, in believing the farmer that no stock was ever going to be in the field or, second, not fencing it anyway.

8.2
Sheep often rub against things, so a beehive is vulnerable if not protected.

From that time onwards all my hives on out apiaries were tied up with a strap right around the hive and hive stand. This arrangement also avoids the hive being blown over in a gale.

There is a much smaller mammal which can cause problems and that is the mouse. Mice are a serious pest of stored combs and may inhabit active honeybee colonies during autumn and winter months. I've only once had a mouse nest in an active colony, and that was before I learnt the importance of mouse guards.

Mouse guards should be fixed across the entrance from late September, but one must remember that the holes through which the bees pass can easily dislodge the pollen from pollen baskets. The guards need to be removed before the spring when the colony begins expanding. Fixing mouse guards in place can be troublesome. If they are nailed in place the bees are sure to come running out with vengeance in mind. You can use drawing pins, but they are difficult to handle with gloves, however thin, although you can get a magnetic tool to hold the pins as you push them in. I never found drawing pins very satisfactory and often found the mouse guard had been dislodged (probably by badgers). I eventually fixed them in place using short screws. These don't disturb the bees as much as using nails, but the bees still come out to see what's going on just outside their front door. Mouse guards cover the full width of the hive, so are clearly not designed to be used in conjunction with an entrance block. I did try using mouse guards with entrance blocks in place and generally it worked well. However, on one occasion at the end of winter I found that one colony had no activity at the entrance, but when

8.3
Hive fitted with a mouse guard.

8.4
This small opening appears to be sufficient to stop mice without the need for a mouse guard.

lifting the roof there was clearly a live, but small, colony there. On investigation I discovered the entrance was blocked by a pile of dead bees. The mouse guard had stopped the bees from easily clearing dead bees away from the entrance, as they would normally have done.

With the advent of open mesh floors, hives have ample ventilation without needing a wide entrance, so I now keep an entrance block, with only a small opening, in place all year round.

I've not used a mouse guard for the last 10 years and have never had a problem with mice in overwintered colonies in that time.

That doesn't mean I've not had problems with mice though, just not in active colonies. I've had them take up residence in empty nuc boxes, and even mini nucs in the shed. They have made nests in my bag of straw smoker fuel. I've had them make a nest on top of the crown board on an overwintering hive that had an ill-fitting roof, or on top of the crownboard in a WBC hive. They have always managed to find ways into what I had thought was a fairly mouse proof stack of supers.

Any hive parts or combs that have been affected by a mouse nest will be avoided by bees as the mice urinate on the combs and woodwork, so any item that has been contaminated by mice needs to be properly decontaminated before reuse, or disposed of.

As mentioned above, some beekeepers may be concerned about badgers. I've seen evidence of scratch marks near the hive entrances, but badgers have never caused any problems with full sized hives. I once had a five frame nuc knocked over and pulled apart, but it wasn't even on a hive stand (my mistake!). If it had been on a stand it might have escaped attention. I have heard of poly nucs damaged by badgers, but they never affected mine as I always sited them on raised stands.

There is a large mammal that is probably the biggest pest to honeybees. It may come as a surprise to many, but beekeepers can be the bees' worst enemy. It should be a rule that we never open a hive of bees unless it's for the benefit of the bees. The pleasure that the beekeeper can get by looking at his bees should never be an excuse for opening a hive! Of course, it is necessary to regularly inspect a hive to assess and control disease or for swarm control purposes, but why should it be necessary to pull the brood frames out on each and every time we need to take the roof of a hive? If we have regularly inspected during the active season, by the end of the summer we will have confidence that the colony is healthy and contains a laying queen. For this reason, after the honey crop is taken off at the end of the year, I never make an inspection below the

"It should be a rule that we never open a hive of bees unless it's for the benefit of the bees. The pleasure that the beekeeper can get by looking at his bees should never be an excuse for opening a hive!"

queen excluder until spring of the following year. The hives will have to be treated for Varroa and fed for the winter, but there is absolutely no excuse for looking into the brood box. I know that many beekeepers, especially those relatively new to the hobby, are unsure of what is going on in the hive and want to reassure themselves by looking. The reality is that opening a hive in September or October and being unable to find the queen will only make the beekeeper more anxious. There is always the chance that when we open and reassemble the hive that we can crush bees, and that risk also includes the queen. Surely, it's better to be unsure about the well-being of the colony, rather than knowing that the colony is queenless, and at this time of year, there is nothing that can be done to rectify the situation. Similarly, opening a hive too early in the year can be potentially disastrous. In this period between the end of August and late March the following year, all we need to do is observe what is happening at the entrance. As a general rule of thumb, if the bees are bringing in pollen when the weather is suitable then they are doing this in order to feed larvae, so things are OK inside the hive. My first inspection in spring consists of removing the roof, and then gently lifting the crown board in order to see how much of the brood box is occupied by bees. Looking further into the brood box will achieve nothing. I made these mistakes in my early years of beekeeping, so can empathise with beginners making the same errors.

There is potentially an even bigger threat posed by humans than by the beekeeper, and that is the risk of theft. It always worried me with out apiaries, that they are vulnerable to theft, particularly hives taken to the moors for heather honey. Throughout my beekeeping life there have been reports of thefts of hives, and it saddens me to know that the perpetrators are beekeepers. I remember in the past we used to have association meetings at members' apiaries. Thinking about this now, I realise that the well-kept secret locations of some of my out apiaries were being shared with all the members of our association. I never lost any hives to thieves, but I know beekeepers that have not been so lucky. When I took hives to the moors, or to fields of borage, I always removed the straps used for travel. Then at least the thieves would have to come prepared with their own straps.

Beekeeping books refer to green woodpeckers as being a problem with over wintering hives. Perhaps I'm lucky but I've seldom even seen one of these beautiful birds. We do have both greater spotted and lesser spotted woodpeckers and neither of these causes any problem. I'm told it is only the green woodpecker that can peck through solid timber. I do recall visiting a beekeeper who lived on the edge of one of the largest raised peat bogs in Britain, an area of special scientific interest. Here green woodpeckers were abundant. He had such a problem that he kept his beehives constantly covered in chicken mesh. I also know of beekeepers who regularly see green woodpeckers, but never have any problems with them pecking hives. In those places where they are a problem they peck through the hive and with their long sticky tongues they can reach in to take the bees. The resulting hole in the hive will let in the cold and damp, so what is left by the woodpecker will probably die anyway. If they are a problem, then wrapping the hive with chicken mesh or black plastic appears to do the trick.

Those are the only warm-blooded creatures that normally might frustrate our beekeeping, although I'm told that pygmy shrews can get into hives and will take bees. Apparently, they leave a pile of empty bee carcases outside the hive.

There are a few insects that can cause problems. The one that comes immediately to mind is the Asian hornet (*Vespa velutina*).

This is an invasive species. At the time of writing, the hornet has not yet become established in mainland Britain, but occasional sightings have been made. It is a big problem in France and has crossed into the Channel Islands. It probably arrived into France in a shipment of pots from Asia. For once, this importation was not a mistake by *beekeepers*. The hornet will take bees from outside the hive and recruit other hornets to that location. Once in the hive they will take live bees and brood. They will also take solitary bees. Between 2016 and 2019, there were a total of 18 confirmed sightings of Asian hornet in England. This figure includes a total of nine nests, all of which were destroyed. It is only time before we will have to deal with this monster. The National Bee Unit produces a very useful leaflet on the Asian hornet. The European hornet (*Vespa crabo*) is not a problem with honeybees although it is a very large insect.

8.5
Asian hornet
(*Vespa velutina*).

8.6
The one wasp that
does cause concern
to beekeepers is the
common wasp
(*Vespula vulgaris*).

Many times, I have heard beekeepers state that they have lost a colony of bees because it was robbed out by wasps. There are a number of potential causes for this, none of which should happen. Like many people I used to regard wasps as undesirable insects, but the longer I kept bees the more I came to realise that wasps play an important part in a balanced eco system. Wasps make beautiful nests from a paper-like material made from chewed up wood. During the summer you often see them nibbling the wood off fencing panels or timber sheds – the reader will have gathered by now that I have an above average number of sheds so provide ample source of wood for wasps. I had an active wasp nest in my store shed all summer in 2020 and they were not a problem. Wasps are carnivores so do a great deal of good in the garden by keeping many garden pests such as greenfly under control. The larvae of wasps produce a sweet substance that the workers take as a reward for feeding the larvae. At the end of the season the colony produces males and finally queens, after which brood production ceases. At this time the worker wasps search out the sweet sugary treats that they are now deprived of. They are great opportunists and will venture into a poorly defended honeybee colony to steal stores, or to feed on the sugar syrup being fed to the bees. If a beehive has a large (full hive width) entrance the wasps will easily be able to sneak in at the edges to rob the hive. Once this robbing starts there appears to be no stopping it. A hive with a small entrance will present a very different proposition though, with the bees able to guard against unwanted guests. Small and very weak colonies are often unable to put up much of a defence however, and often get completely robbed out. Large healthy colonies with a small entrance are unaffected, but colonies that often are too small and weak to survive the winter die out. It is these poor colonies that many beekeepers say they lost because they were killed by wasps. Small colonies are quite capable of resisting wasps if the entrance is small enough and care is taken when feeding that no syrup is spilled.

Another insect that many beekeepers encounter is the wax moth. There are two sorts – the greater and the lesser wax moths (see Appendix 4). These are generally not a problem in large active colonies, but I have seen them in old hives that have a pile of empty supers on the top of a small struggling colony. Unfortunately, there are far too many of these almost abandoned hives around, being given little attention by the owner. The one time the wax moth will become a problem for beekeepers in general is supers in storage. There are treatments available, but they only work during certain periods of the lifecycle of the moth. There is, however, one treatment that is available and free to all. The moth larvae do not survive freezing! If you store your supers in a position where they are exposed to harsh winter conditions the moth is normally kept under control, so it is better to stack them outside than to put them in a frost-free shed. As they don't like honey, storing supers wet after extraction will limit the damage done by the moth. Before I learnt these little tricks,

I lost many good quality super combs, and brood combs as well! With climate change, long cold, sub-zero, winter days may become a thing of the past and then we might have to resort to some other method of control.

While working as an inspector, on two occasions, I came across bee-keepers using moth balls (naphthalene) to control wax moth in stored super combs. Naphthalene is highly toxic to both humans and bees and will also leave residues in honey. I've no idea whether they stopped the practice once I pointed out their mistake, but they had been doing it for years. I don't think you can buy moth balls in shops anymore, but like most things you can get them online from the Far East.

The Varroa mite was absent in my early years of beekeeping, but initially caused widespread problems when it arrived in this country, and we are still learning about it and how to mitigate its effects. One feature we noticed with hives that had a high level of mite infestation was bees with stunted wings. It was thought at the time to be due to bees being damaged by mites during development but is actually due to a virus transmitted by the mite, called stunted wing virus. No one knew then

8.7
Varroa mite on a honeybee.

the extent of the problems caused by this mite and as beekeepers we are still learning. I remember in those early years not wanting to use unnecessary chemicals in the hive. There were many homemade treatments tried by almost everyone. Two licensed treatments were available, Bayverol and Apistan, which both contain a synthetic pyrethroid. I was very reluctant to put these products into my hives, due to concern over chemical residues in the honey, but that was probably a mistake. There was lots of talk about using Thymol, which is a synthetic thyme extract, it may have kept mite levels down, but not sufficiently to control the virus impact that is associated with Varroa. I probably lost half my colonies over a period of three years. Reluctantly I resorted to using Apistan strips and colony losses dropped dramatically to only about 15% per year. The important thing when using any medicine is to do exactly what it says on the instructions. Long-term use of these treatments resulted in the mites becoming resistant, and perhaps surprisingly, overdosing, or over prolonged dosing, accelerated the mites becoming resistant. I remember as a bee inspector testing colonies for resistance. I met beekeepers who put Bayverol strips in every year in autumn and rather than leave them in for the prescribed 28 days took them out the following year! One beekeeper I met had been doing this for years and said that they didn't think it was necessary as they didn't have Varroa, as they had never seen it. That was the only occasion that I actually saw Varroa mites running on the face of the combs. (Another example of when a good pair of glasses might help.)

I've seen colonies that had been treated in the autumn that collapsed before they reached the spring, until it was realised how important it was to reduce the mite load before the time that the colonies raised their winter workers.

I've also seen colonies that have very low mite numbers where stunted wing virus is still present. There was much to learn, and we are still learning!

When the mites became resistant to Apistan and Bayverol we had to change to different type of system. Elsewhere in the world beekeepers have used other more toxic mite treatments, but in the UK the treatment that most beekeepers changed to then was Apiguard. This treatment, which is a special formulation using Thymol, can be up to 90% effective if applied correctly. Not as effective as the 98% with Apistan when it worked OK. This of course is when things got difficult again. Let us assume you have 2000 mites, a 90% kill will still leave 200, and these will double in number about every 3 weeks. If for example you treat on 1 September, by the third week in the month the mite numbers will be 400, and by mid-October there will be 800. If the weather is mild and brood production continues then by early November there will be 1600 mites. The bees raised at that time are expected to last through the winter but will have been compromised during their development by the mites.

This was almost a perfect storm situation, we had lost the single most effective treatment due to resistance, and the alternative didn't quite do enough for us to just get away with a single annual treatment. Just when we thought that beekeeping was getting difficult it had got a whole lot worse. No wonder older beekeepers gave up and new beekeepers just couldn't get to grips with the problem.

I dislike the idea of using more and more different toxic chemicals. As one treatment becomes less effective another one takes its place. Another widespread treatment is sold under the name Apitraz, the active ingredient is Amitraz (a widely used agri-insecticide). This treatment is 99% effective, but as with previous treatments the mites will develop resistance.

For a few years now I've been using MAQS (mite away quick strips). These strips use formic acid, which can penetrate the cappings of cells and kill mites everywhere in the hive, including the 80% of mites that are in the brood cells. Treatment period is only 7 days, compared with 28 days for other types of treatment. It is about 90% effective but there is anecdotal evidence that queen loss may occur. I occasionally lose a colony over winter due to queen loss, but I don't know whether this is due to the mite treatment or some other cause of queen failure. Be sure to read the instructions before use. It is not suitable for use on small colonies, or nucs. During treatment daytime temperatures should be above 10°C, but it should also not be used in very hot daytime temperatures, above 25°C, as there may be significant brood loss or even queen loss.

Monitoring of Varroa mite levels is important, but as it is time consuming, I understand why many beekeepers don't bother. It is very easy to assume that all the hives in one apiary have the same levels of mites, and this is a trap that is very easy to fall into.

One year I closely monitored only one hive on each of my apiary sites in order to reduce the size of the task of checking mite levels. I was surprised to discover at the end of the season that mite levels varied widely across the apiary. The lesson I learnt was that if you are going to monitor mite levels then you have to do it on all hives, not just a few colonies.

It is clear to me that some colonies deal better with the mites than others. Some have a high mite level and a low incidence of apparent virus problems; others have a very low mite count, but have a high level of virus related issues. We need to adopt an integrated management plan (IMP). Some years ago, I changed all my floors to open mesh so that I could check mite drop, but now I seldom check natural mite drop. However, the open mesh floors remain open all year round as part of my IMP. I've heard that an open mesh floor could reduce mite numbers by about 10% as mites that fall through can't climb back into the hive provided the clear gap under the hive is more than about five centimetres.

A word of warning about open mesh floors! If you leave the monitoring tray in situ, it collects debris that drops out of the hive. This is mostly wax cappings or to give them another name 'food for wax moths'. You will have provided an ideal wax moth breeding place, so only use the insert when you want to monitor the mite drop.

I've sourced a number of different open mesh floors over the years and converted some solid floors to open mesh. They all have slightly different size insert sheets, so that when I come to use them, I can never find the right one.

Some years ago, I started doing a winter treatment using oxalic acid in addition to the normal late summer treatment. This acid works by burning the mouth parts of the mite but is harmless to bees if administered correctly. It does not penetrate through the cell cappings, so is really only effective at a time when there is no brood. The one time when we can be sure there is little or no brood is in the middle of winter. At this time any mites will be on the bees and are vulnerable to acid treatment. Of course, we don't want to open the hive up in the middle of winter to check if there is any brood, we just have to assume there is none. For some years the recommendation was to do the treatment at the beginning of January, but I'm sure that is a mistake as some colonies may already have a small amount of brood and any mites will already be in that brood. I now do my winter treatment in mid-December as I think it more likely that they will be broodless.

Until recently, oxalic acid dihydrate was a non-licensed treatment that many beekeepers used, but recently some companies have licensed it. Now we pay more for a little packet of oxalic acid crystals, with instructions how to use it, but at least we can do it within the rules. There are two main ways of applying the treatment. The first, which I use, is referred to as the trickle method. The oxalic acid is mixed in accordance with the instructions in a sugar solution and then trickled down between the frames on top of the winter cluster. The required dose is 5 ml per seam of bees. The cluster normally occupies five or six seams, so the dose is 25 ml or 30 ml per hive. Experimentation was the name of the game in the early days, and I thought 5 ml per gap between frames would mean a dose of 50 ml per hive, whatever the size. This was too much, particularly for small colonies or nucs. I have to admit that I probably killed a few through overdosing. Now I tend to under-dose rather than over-dose.

The other way to use oxalic acid is to use an evaporator. I've always thought that sublimating an organic acid was a potential health hazard, so have steered clear of this method. The trickle method is quicker and just as effective. Whichever method of application is used the effectiveness is about 90% to 95%.

There is one mistake, that I keep coming across, when using the trickle method. The important thing to remember is to keep the disturbance of the winter cluster as little as possible. In order to do this, I remove the hive roof and prepare the dosing syringe before gently lifting the crown board and assessing the number of seams of bees. The dose is then quickly administered before replacing the crown board and roof. Those beekeepers who have colonies on brood and a half are often in a dilemma about how to do the treatment. So often I have heard of beekeepers removing the 'half' to dribble the solution onto the bees in the bottom brood box. This normally will mean that the brood cluster is split in two and on reassembly there is inevitably the risk of squashing bees (or even the queen). My advice is always to never take the brood and a half apart to do the treatment. Just treat the brood box as if it were a single box. It may be necessary to use a torch to shine down between the frames to see where the bees are!

To summarise my current Varroa strategy, I do my main treatment at the end of August (using MAQS) followed up by oxalic acid treatment mid-December.

I monitor mite numbers during normal hive inspections by inspection of drone pupae.

If treatment is necessary during the active season then the choice of treatment would depend on weather, temperature and state of the colony. Apistan or Bayverol strips can still be useful in this situation if they've not been used for 5 or 6 years, as they will still give 90% treatment. They cannot be used again for a further few years due to the high levels of mite resistance. My overall mite control success has reduced normal winter losses to about 10%. I've not noticed a mite on either a bee or on the comb in years, and I can't remember when I last saw deformed wings. Monitoring of the drone comb now normally only shows a small number of mites.

Many beekeeping books mention the *Acarapis woodi* or tracea mite. Over the years I've looked for these mites but never found any. That is probably because treatments for Varroa will also be effective against the tracea mite. I've done microscopic dissection of worker bees and failed to fine any infested tracea. It was once widely thought that Isle of Wight disease was Acarine (the infestation of *Acarapis woodi* mite), but this has now been largely discredited. The signs of the disease appear to match that of chronic bee paralysis virus, which may very well have been transmitted by the Acarapis mite. Isle of Wight disease caused widespread colony losses, to the extent that some writers later stated that it wiped out the native strain of bee. I think we can safely say that *Apis mellifera mellifera* is still 'alive and well' and living in quite large pockets of the UK. Unfortunately, in an attempt to make good the colony losses, bees were imported in large numbers from Italy, resulting in widespread mongrelisation of our honeybees.

8.8
Inserting an uncapping fork to test drone cells for Varroa.

8.9
This picture shows exposed drone pupae with no sign of Varroa mites.

8.10
This picture shows just one mite on ten pupae, indicating a low level of mite infection.

There are other bacterial diseases that we need to be vigilant about. European foul brood (EFB) is widespread in many parts of the country, and at one time was controlled by the use of the antibiotic oxytetracycline hydrochloride (OTC). While this may have helped to control the symptoms, it didn't actually control the disease, so it is not the treatment of choice, although it is still an option. Normal control now is by using a shook swarm. I once had EFB, it was found by the bee inspector checking my bees because a neighbouring beekeeper had a case. It was possibly transmitted by drones drifting between hives. We treated using a shook swarm, and OTC which was still being advocated at that time. I never saw another case in my bees. Often EFB will crop up unexpectedly either at the start of the season or at the end. Well-fed colonies often show no signs of the disease. EFB is far more difficult to eliminate than American foul brood (AFB), which has far fewer cases. AFB is an easier disease to treat, and to stop it in its tracks, provided it is identified before the colony collapses. Treatment is by destruction, under the supervision of the bee inspector; all the combs and bees are destroyed by fire and the hives sterilised using a blowlamp. The disease is caused by a spore forming bacteria and the spores can live for many years in old equipment. It can be contained by careful management of your equipment, particularly hive parts. I've seen cases where beekeepers have bought second-hand hives and these hives have brought the disease with them. For this reason, it is necessary to sterilise all hive parts before putting them into use. Wooden hives are easy to sterilise using a blowlamp – just scorch all the timber to a light chocolate brown. Old frames should just be disposed of by burning if possible. When I was an inspector, I found it so frustrating when dealing with one particular outbreak of AFB where the beekeeper had multiple apiary sites. The disease kept on cropping up because supers were being transferred between different apiaries. We've probably all done that, but when you have a case of AFB you really do have to go back to basics as far as hygiene is concerned.

With all the emphasis in the last few decades on Varroa what happened to discussion on Nosema? All the old books on bee diseases mention Nosema disease and its causative organism *Nosema apis*. There is now a new variant *Nosema cerana*, which has come from the Asian honeybee; it is only distinguishable to the trained eye from *Nosema apis* under a high-powered microscope, so this is outside the scope of the average beekeeper. In all my time as a beekeeper I have only attended two association talks on Nosema, and one of them was a talk I gave! This disease is largely unknown and mostly unspoken about, but it is probably more widespread than we believe. It is likely that it is one of the major causes of winter losses after Varroa. One symptom of Nosema is dysentery, but just to add a complication *Nosema cerana* does not appear to show dysentery. So, what can the average beekeeper do? It used to be treated with Fumidil B, which is an antimicrobial agent, but the marketing authorisation for Fumidil B expired on the 31 December 2011. Any existing stocks of the product can be used up until the end of the expiry date shown on the packaging, which will have long past. This leaves us with no approved treatment. (Although it is still listed on Amazon and eBay, it would be a mistake to import and use it.) Nosema can be controlled by changing the bees onto fresh clean combs, by a Bailey comb change (see Appendix 5), the idea being that the Nosema spores will be left on the old combs. Alternatively, you could do a shook swarm (see Appendix 6), but this is not as gentle on the bees as the Bailey comb change.

In conclusion, when I used to teach beginners, I always kept discussion on diseases until the very last, thinking that it's best to get them hooked before actually telling them about the difficulties. The main things we need to be able to do as far as diseases are concerned is treat Varroa as required, keep Nosema under control by regular comb changes and apply good apiary hygiene. If we do all that what else could go wrong? There is climate change, loss of biodiversity and, of course, the small hive beetle, which is presently confined to Africa, North America, Australia, the Philippines and a small area of southern Italy. If the small hive beetle arrives in this country it is likely to be spread by movement of queens, packages of bees, honeybee colonies, honeybee swarms, honeycomb, beeswax, beekeeping equipment, soil and fruit, or movement of alternative hosts (for example, bumble bees). Therefore, while we continue to import queens from Italy there is a real and present risk that the beetle will get here eventually.

9

The honey harvest

There is something satisfying about producing your own food, whether it is growing your own fruit or vegetables or harvesting your own honey. Many hobby beekeepers only want to produce sufficient for their own needs, although it is surprising how many members of the family and friends discover a liking for honey once you become a beekeeper.

9.1
'Get that out of my kitchen!'

The first problem the new beekeeper has is getting the honey off the hive. The simple advice in the books was to put a Porter bee escape in each of the holes in the crown board and put this under the supers to be taken off, and then leave for 24 hours. With my original old WBC hive this proved to be woefully inadequate advice. The design of the boxes in a WBC relies on the use of metal or plastic end spacers to make them 'bee tight', and if you use your crown board as a clearer board then what do you put on the top to keep the bees from going in at the top? My first attempt to clear the supers was a failure – after 24 hours there were just as many bees in the super as there had been the day before. I found that the only way to make it work was to have a second crown board, and to clear just one super at a time. Another problem was the Porter bee escapes. The little springs can become broken, bent or incorrectly adjusted so they don't function as a one-way valve like they should. If the escape has been left in place in the crown board for a lengthy period of time the bees will have glued it up with propolis, and then when you need it to work as a clearer board no bees can get through. Often some drones have managed to get into the supers, and these are then too big to pass through the Porter bee escape and they become stuck causing a traffic jam.

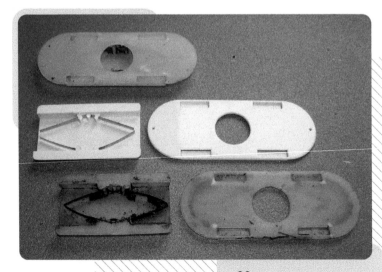

9.2
Porter bee escapes.

Things are a little easier with the National hives that I soon changed to, as the crown board can be used without needing another cover board on top as the roof is a bee tight cover. Unfortunately, there are still things that can go wrong. One of the first things I found with my rather old, acquired second-hand hives was that the ventilation holes in the roof often didn't have mesh covering them, or in one case the roof had been put together wrongly so that the mesh didn't actually work. These issues resulted in bees, and/or wasps being able to gain access to the top of the hive. Other problems with old equipment are that there may be parts that are split, have knot holes or just have wear and tear on the corners of the boxes meaning that the supers are not bee tight (sealed so that bees can't get in). The result of these defects is that when leaving the clearer board on for 24 hours the super can be robbed out by bees or wasps.

There are various alternative clearer boards that don't rely on mechanical springs to prevent bees returning to the supers. The first of these is the Canadian clearer board. There are various designs of these boards, but they all use cone escapes rather than one-way valves. The bees clear rapidly out of the supers and find it difficult to learn how to get back through the cones. The Canadian clearer board usually clears the supers in a couple of hours.

There are a number of other bee escapes which are worthy of mention. The first of these is the circular or eight-way escape, and the second is the diamond or rhombus escape. These have no valves, but rely on the bees clearing the super into the edges of the brood box. When trying to go back they are initially attracted to the centre of the board so don't find an easy way back into the super. These escapes rapidly clear the bees in a couple of hours. The last type of clearer board that I use is based on a triangular mesh pattern that I found online some years ago. I know at least one of the UK equipment suppliers now sell a clearer board to this design and it is called a Forest board. I decided to make some of these myself, but rather than triangular mine are square (basically because I found it easier to make a square one than a triangular one). These boards work the fastest of all and clear about 95% of all the bees from a single super in less than an hour.

9.3
Rapid clearer board.

There are two other ways of clearing bees from the supers, neither of which I have used. The first is using a fume pad and a bee repellent and the second, used by some commercial beekeepers, is a bee blower – used to blow bees out of the super.

To recap, having put on our clearer board, we can then return after the required time and take the supers off without disturbing the bees. It doesn't always work that easily though. With a very large colony there normally isn't enough space in the brood box to accommodate all the bees – they simply can't go through the clearer board, of whatever sort, because there is a logjam. It is best if we can put an empty super underneath the clearer board to allow the bees some cluster space. With a homemade clearer board it is possible to incorporate a deep gap which provides the cluster space. The bees are often less inclined to clear from frames of unsealed honey, whereas they readily leave sealed honey.

My first attempts at extracting with an old tinplate hand cranked extractor were interesting to say the least. The kitchen is the obvious place to extract your honey! Only a few bees remained in the super and they soon flew to the window (having flown round the room a few times first, much to the consternation of my wife and children). We set about uncapping the frames, trying not to let the cappings fall onto the floor (some did of course). The first two frames were put into the extractor and we all took it in turns to wind the handle. The excitement soon waned and younger assistants disappeared. This process was repeated until all the honey was extracted and the honey collected in buckets to be transferred to the honey tank. At this point one realises that there are bits of honey and beeswax cappings on the floor, which by now have been transferred to most rooms on the ground floor of the house. It is surprising how a small drop of honey on one's hands ends up on every door handle in the house. It takes considerably longer to clear up the kitchen (and the rest of the house) than it does to extract the honey. Next problem is how to get the propolis off the work surface!

After a couple of seasons of extracting in the kitchen it was decreed that extraction should take place in the utility room, and at that time I acquired an electric ten-frame extractor. That system worked quite well but the problem of sticky door handles remained! There is always the problem of bees when extracting. Not only are there the bees which are still in the supers, which you bring into the extraction room, but as soon as you start extracting you attract hundreds of other bees. Unless your extraction room is absolutely bee proof they will get in, normally under the door, or through gaps around the edges. I once saw an extraction room where the door had draft excluders all round, but it was an old building with a large old lock and the bees got in through the keyhole. Of course, every time you open the door bees get in. Bees are attracted to the house by the aroma of the honey, so there is generally a gathering just outside the back door. If you only have a small amount to extract, then try to do it in the evening or on a wet day.

When my number of hives really expanded, I was finally banned from the utility room and set up a purpose-built honey extraction and bottling facility at the far end of the garden – shed number 2! It was rather

"It is surprising how a small drop of honey on one's hands ends up on every door handle in the house."

interesting when I had my first inspection by local authority environmental health department. I took the lady inspector down the garden to my 'shed', which was fully equipped with washable painted walls, work surfaces and splash backs with stainless steel double sink with hot and cold water. It also has an electric insect killer, which interestingly attracts and kills wasps but not honeybees. At the end of the inspection I was asked if I was happy if she took some photographs. Initially I was a bit taken aback, but apparently, they had no pictures of best practice. For once I had managed to get it right. A few days later my five-star food hygiene rating arrived in the post. I know of one commercial beekeeper in a different county who was inspected by their local authority, and one of the inspector's comments was 'I'm afraid this won't do – there are signs of insects in this room.' It does sum up the lack of understanding about honey by the lay person. When I had my first visit by our local trading standards officer the first question I was asked was, 'Where do you source the ingredients for your honey?'

One of the most useful pieces of beekeeping equipment during extraction is a tidy tray, obtainable from most garden centres. Just the right size to fit a super! You can use these to transport supers in the car to avoid drips of honey. I use them in the extraction room to stack supers and on the work surfaces to avoid honey and propolis.

9.4
Tidy tray with a super.

What do we do with the supers now they have been extracted? There are two options. The first is to store them dry, and to do that we put the supers back on the hive and let the bees clean them up. If you do this, you must put the supers back on the hives in the evening to reduce robbing! Then of course we have to clear the bees back out of the clean supers. The alternative is to store them wet. This has the advantage that they are less likely to attract wax moths when they are wet. The honey left on the combs may ferment, but when they are put back on the hives in spring the bees very rapidly go into them. As far as actually storing them, ideally, they should be somewhere that is exposed to frost, because that will kill wax moths. When I had 60 hives and about 150 supers, I stacked them outside with a sealed hive floor underneath each stack and a hive roof on top. If you don't ensure the stack is absolutely sealed up don't be surprised to find a mouse takes up residence during the winter! I've lost track of how many mice I have given home to over the years.

I've always uncapped my combs without using any heat. In the early years I used a simple piece of wood with a recess to hold the bottom of the frame when uncapping. When I increased my number of hives, I used a large cold uncapping tray.

9.5
Cold uncapping tray.

9.6
Heated uncapping tray.

Many beekeepers use a heated uncapping tray, which has a water bath and a heating element. Some beekeepers use this water bath to heat their uncapping knife. I was unaware of that trick, but when visiting one beekeeper for his general husbandry assessment he pointed out that the tray he was using belonged to his beekeeping association and the water bath had mould on the inside and was virtually impossible to clean. That was an eye opener, particularly if used to heat the uncapping knife. The one advantage of the heated tray is that it melts the cappings and that makes handling that material much easier, but you do risk overheating the honey.

Once the extraction has been completed the extractor will have to be cleaned. Initially wash everything in cold water. This might sound strange advice as normally we wash things in hot soapy water. Unfortunately, propolis and wax melt, so using hot water will melt them onto the surface where they will be almost impossible to remove. Once you have washed all the honey and wax away using cold water then you can use hot.

The next job after you have extracted your honey, and cleaned up all the sticky mess, is to bottle it, but the honey straight out of the extractor still needs a lot of effort before you can put it into jars. One thing which is not really apparent, unless you are using a heated uncapping tray is how much honey there is left on the cappings. I think it's about 10% of the honey crop. My method for dealing with the honey and the cappings is to strain them all though a fine conical sieve. The honey straight from the extractor is put into a bucket for transfer to the strainer.

Once all the honey has been strained, I then put the cappings into the same strainer and leave it for 24 hours. There is very little

9.7
Conical strainer.

honey left on the cappings at this stage, but some people carefully wash them and use the resulting sugary liquid to make mead.

Once filtered, depending on the quantity of honey, if it's more than the 50 kg capacity of my honey tank, I'll put it into 15 kg buckets with airtight lids for storage. A word of warning at this point. Leave only a very small air gap at the top of the bucket, or the honey may absorb water and start to ferment. If it does it will foam over the top and make a mess. The honey will no longer meet the requirements of the honey regulations for sale.

Honey has a beautiful aroma and a delicate taste, which varies depending on the flowers from which the bees have collected nectar. The last thing we want to do is destroy these qualities of taste and smell. Unfortunately, that is exactly what happens with mass produced honey. I've lost count of how many people have told me they don't like honey. Once they taste a local beekeeper's honey they are often converted.

Until I made a proper thermostatically controlled warming cabinet, I struggled to melt solid honey without overheating it. For many years I used an insulated box with two 100 W tungsten filament electric light bulbs. You can't even buy these bulbs now. My warming cabinet now has two oil filled tube heaters and is thermostatically controlled. I also use a freezer thermometer to check the temperature (it works at hot temperatures as well as cold and has a remote sensor). The cabinet was designed to fit three 15 kg buckets but is also useful for warming jars of honey.

Honey from different sources needs to be treated differently. Some honeys set very rapidly, whereas others set very slowly. I feel myself lucky that living halfway up a mountain in North Wales there is no oil seed rape (OSR) within flying distance, but I'm only 1 mile (1.6 km) away from the heather moors. For some beekeepers the OSR is their major honey source. I've taken bees to the OSR in the past and it isn't my favourite honey. It has a very high glucose content, so granulates very rapidly and with a very fine granule. As soon as the flowers start to fade the beekeeper has very few days in which to get the honey extracted. If it sets in the comb the only way to deal with it is to cut it out of the frames and melt it on a heated uncapping tray. As I don't have a heated tray I had to improvise by cutting it into chunks and putting into buckets and melting it (gently), then mashing it all up before putting it through a strainer. Then after all that messing about, I'm not keen on the taste anyway. It is useful though to blend with another strong flavoured honey to make into a soft set honey.

Heather honey has problems with extracting. I found the best way to deal with it is to cut it up as cut comb pieces, those bits left that aren't suitable can then be pressed to extract the remaining honey.

Some years ago, I made the mistake of lending my heather press to another beekeeper and didn't have any heather honey that needed extracting for a couple of years. When I came to need it, I asked the beekeeper who I thought had borrowed it if I could have it back. He denied ever having it! No one else I could think of owned up to having it either! I'm reminded of that quote from Shakespeare's Hamlet 'Neither a borrower nor a lender be; / For loan oft loses both itself and friend.' I had to buy another cheap press, but it doesn't work as well as the old one.

Heather honey has particularly strange characteristics in that it is thixotropic; that is, it is runny if stirred up, but then partially solidifies

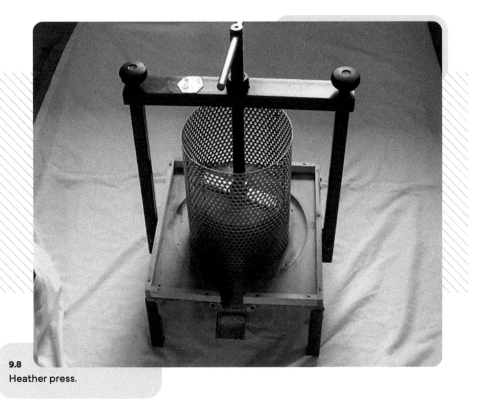

9.8
Heather press.

into a jelly, rather like non-drip paint. In order to filter it and bottle it you have to warm it then stir it up. It took years before I got it right, but initially I couldn't filter it, I sometimes overheated it and I made some awful sticky messes. The only thing left to do, with one part-bucket full, was to make mead with it. The good thing was that 14 years later the mead won first prize in the association honey show. Another thing that I managed to get right by accident!

When you first have some honey, it is very tempting to give it away to family and friends. My naivety about bees extended to jars for honey. I never knew what a great variety of shapes and sizes of jar there is. Fortunately for me there is a good supplier of beekeeping equipment and glass jars nearby and I realised I wasn't restricted to the old style 1 lb (454 g) honey jar. Until fairly recently honey had to be sold in 1 lb (454 g), 12 oz (340 g), 8 oz (226 g) or 4 oz (113 g) jars, now it can be sold in any size container. If you must give away this hard-earned delicacy get some smaller jars!

When I had lots of honey to sell, I wanted my jars to be different, so that they stood out from the rest. There was also a problem with the lids on the old screw top honey jars; when lifting the jar by the lid the lid came off, fortunately this problem has since been resolved. Modern jars have a twist lid rather than a screw top. I chose a 1 lb jar with a honeycomb pattern on the top edge. For some years I just filled the jars until they looked full, until one day trading standards insisted that I get some proper scales. When I filled my jars up as usual and used my new weighing scales, I found that I had been putting an extra ounce of honey in each jar for 2 years, which amounted to about 25p given away with each jar. The new scales had cost £200, but that outlay was paid for in the first 3 months of use.

Leave the honey to settle for a few days before running it into jars. If you try to bottle it without leaving it to settle all your jars will have an unattractive bubbly, scummy layer at the top. Hexagonal jars look attractive, but they are difficult to fill, as a small bubble is left in each top corner of the jar, which has to be removed with a thin bent rod (a grapefruit knife works well).

It took me years to master the production of soft set honey. Unless you have thermostatically controlled heating and mechanical stirring on your honey tank it can be tricky. My first attempt at set honey was simply to allow the honey to crystallise in the jar, and with a honey like OSR the result is a jar of almost rock-solid set honey. The second attempt was to blend the OSR with honey from another source that wouldn't set quite as rapidly and the result was almost the same, but granulation took slightly longer. The guidance I had been given on producing soft

"For some years I just filled the jars until they looked full, until one day trading standards insisted that I get some proper scales."

set honey is to warm a bucket of set honey to partially soften it, then mix it with some liquid honey and then stir the two together, while avoiding getting air bubbles in it. Leave it to partially set, mix again and then bottle. If it is not left long enough it will be too liquid, left too long it won't run through the tap into the jar. I once left it too long and I couldn't get it out of the tank. I had no way to warm it, as my warming cabinet wasn't big enough to take a 70 kg tank. The only way to re-warm the tank was to put it in the bath and partly fill the bath with hot water. Carrying a 70 kg tank upstairs to put it into the bathroom was no mean feat! I realised at that point that if I was going to produce a consistent soft set honey in volume, I needed a proper creamer. I was producing about 1 tonne of honey per year, and half of it was being sold as soft set, so I invested in a motorised creamer with a thermostatic temperature control. From now on soft set honey was easy. With about 15 kg of set honey in the tank a further 30 kg of runny honey was poured in, then it was just left to mix together without any heat. Once mixed the honey was then left to set. Each day the honey was mixed for a few minutes, until the required thickness was achieved. At this point the temperature was raised to about 28°C and mixed again. Keeping the honey at that temperature it was then bottled – resulting in perfect soft set honey. I would bottle 30 kg of honey and then top up the tank with 30 kg of runny honey for the next batch. This system worked most of the time except in very hot weather, when my honey room (shed no. 2) became very warm and all the honey in the tank fully melted. I had to swap a bucket of my beautiful mixed floral honey for a bucket of bland tasting OSR honey with another beekeeper to get the whole process restarted.

There are many styles of label; you can buy them printed to your requirements or you can print them yourself if you have a printer. So many jars you see from local producers don't have all the required information to comply with the honey regulations, but if you are selling direct to the consumer you don't need to have all the specified information on the label. One of the requirements is to have a lot number and a best before date. I recall a discussion a few years ago with trading standards about what would be an appropriate timescale for a best before, as I had heard that honey had been found in the ancient tombs in Egypt which was still edible. I was told that it was up to me to state a suitable date, but I doubt whether 2000 hence years would have been OK. I decided that just a few years would be acceptable, and I remember putting best before end 2010. I had a phone call a few weeks later, from a customer who had purchased my honey from a local shop. The lady was concerned because the label stated, 'best before end 2010' and as it was now 22 October, 'Would the honey still be OK to eat?'

If the best before date is specific and can relate to the date the batch was produced, then it doesn't require a lot number. I specify 4 years from the date of bottling, others may choose a different length of time, and with a day and date I no longer need a lot number, or the register you need to keep of the particular batch associated with that lot number. This requirement is all to do with product recall, but frankly I don't know of any problems with local honey that would merit a recall. I do recall honey from non-EU countries being taken off the supermarket shelves some years ago.

The Honey (England) Regulations 2003 specify the maximum water content of honey. In general honey must not exceed 20% water content, except for heather honey which should be no more that 23%. The reason for these limits is that honey with high water content will ferment. Normally a mixed floral honey will have water content much lower than the stated limits, generally being about 17%. Most beekeepers have no idea of the water content of their honey, and no way of checking it! I suspect the reason for that is simply cost. To be able to measure the water content you need a refractometer, which will cost around £50 (or £180 for a digital one). Even if you do have a refractometer, what are you going to do if your water content is too high?

Let us first address the possible cause of high water content. Honey is deliquescent, that is, it absorbs water from the air. Bees stop this from happening by sealing over the cells of stored honey with wax; however, any cells that have not been filled will remain uncapped. Towards the end of the summer the nectar supply runs out, unless you are in a location that has a late crop, such as heather or Himalayan balsam. We have probably all made the mistake of leaving the supers on just in case there is a bit more honey to be had. From my experience this never happens! The longer we leave it the more uncapped honey there is.

The beekeeper can then make things worse by not extracting it as soon as the supers have been removed. These uncapped cells contain honey that has not been fully processed by the bees to reduce the water content. If the supers are left in a cold and damp storage location the unsealed honey will pick up even more water. If extracted honey has only been from sealed comb the water content should be about 17%, but the larger the amount of unsealed comb that is extracted the higher the water content will be. So what? Does it make any real difference? Unfortunately, it is important. First, it won't comply with the regulations but, second, the honey will probably ferment in the jar. Fermentation doesn't happen with liquid honey in the jar, but it does with granulated honey. The water content of honey crystals is lower than the water content of liquid, so as honey granulates, the water content of the remaining liquid becomes higher. This is not a problem with honey that has a low water content as the remaining liquid between the crystals will still have a relatively low water content. With a high water content honey, when it granulates, this liquid between the crystals will have even higher water content and will ferment due to the wild yeasts in the honey. As the honey ferments, the CO_2 (carbon dioxide) produced by the fermentation process causes the honey to foam and expand and it will overflow out of the top of the jar! A jar of liquid honey can sit on the shelf in a cupboard for months, during which time it will granulate, then if it ferments it starts to overflow and causes a sticky mess on the shelf. In this state the honey should only be used for cooking. In addition, honey can't be sold if it has begun to ferment or has fermented, so it's better to avoid the problem in the first place.

I know of one commercial beekeeper who uses an industrial dehumidifier in his extraction room to ensure the water content of his honey is kept as low as possible.

It is important to check the water content, but what can you do if the water content is too high? Commercially it can be dried using specialist equipment. It is possible to reduce the water content by using a warming cabinet at about 30°C and a dehumidifier to slowly reduce the water content, but if you raise the temperature too high then you will spoil the honey. Overheating or keeping the honey at an elevated temperature too long will raise the hydroxymethylfurfural (HMF) content above the permitted 40 mg/kg and will remove the entire wonderful aroma and taste from your honey. You might be able to blend different batches of honey to achieve the required water content, or as a last resort use it to make mead.

Heather honey has higher permitted water content of 23%. Heather honey I have produced is normally lower than this, but much of the crop comes back from the moor in late September, when the temperatures at night are cool, and is uncapped, so can quickly pick up water. Heather honey doesn't granulate though. As it is thixotropic it is difficult to extract, but it doesn't granulate, so even at 23% water content it doesn't ferment.

As I noted earlier when storing honey, it is imperative that it is stored in sealed containers with as little air gap above the honey as possible. This will limit its ability to pick up moisture. This is also true of honey in jars. The jar I use is nominally 12 oz or 340 g, but when filled with this exact weight the honey only just comes up to the shoulder of the jar. I put in 370 g of honey, which makes the jar look full and reduces the head space above the honey. With the latest regulations on permitted sizes we can sell in whatever size we want – just adjust the selling price to reflect the quantity in the jar.

Beeswax is often a forgotten hive product, because beekeeping books don't really cover this in any depth.

Each year a colony of bees will produce about 0.5 kg of beeswax, but the collection, filtration and use of this product of the hive is so often overlooked. The harvesting and processing of this valuable product appears to be shrouded in mystery. *Beeswax: Production, Harvesting, Processing, and Products* by William L. Coggshall and Roger A. Morse was published in 1896, and it wasn't until 1981 that Ron Brown published *Beeswax*, and this book is still in print today. No doubt processing beeswax on an industrial scale is relatively straightforward, but on a small scale it can be a messy and frustrating business.

My first attempt to melt beeswax was using a homemade solar extractor. This worked reasonably well, but required bright sunny days to melt wax successfully. It also needed to be regularly turned around to face the sun. It would also attract lots of inquisitive bees. It was fine for melting a few odd bits of wax or even a comb from a super.

Dealing with combs from the brood box it is a whole different ball game. If you try to melt a brood comb in a solar wax extractor, the paper-like linings of the cells will absorb the wax. One of the first beekeeping books I read described putting the combs in a hessian sack and immersing them in a water filled metal drum over a fire. I don't know if it would have worked, as I don't have the space in the garden for a roaring fire and an old oil drum. Dealing with old brood combs remained a mystery for many years until I learnt about steam extraction.

9.9
Homemade steam wax extractor.

This method uses an old solid floor, a brood box, a mesh screen and an old hive roof fitted with a pipe to attach it to a steam generator (wallpaper stripper). It might look a mess, but it's just made from bits that are fit for nothing else. It works reasonably efficiently and will extract a small amount of wax from brood combs in about 30 minutes. As with the solar extractor it is a focus point for my bees, which are only a few yards away from my bee shed/workshop. I find it best to keep my wax extraction to the winter months to avoid this bee problem. It also is an ideal job for this quiet time of the year when there is so little beekeeping to be done. The resulting wax is relatively clean and is generally a pleasant colour. It is all too easy to produce rather grey coloured beeswax, rather than a pleasant golden cream colour.

Cappings can be melted in the steam extractor, although that is not the way I normally deal with them. My honey warming cabinet will heat up to 65°C, but obviously this is too high a temperature for honey but is around the melting point of beeswax. All my cappings are just put into buckets and placed in the warming cabinet. Once the wax has all melted (about 24 hours) the cabinet is turned off and the wax left to cool. The resulting wax slab is taken out once it has gone hard and the dark brown sludge, or slumgum, is scraped of the underside of the block.

If you want to filter wax it is virtually impossible putting it through any sort of straining cloth as it solidifies as soon as the temperature drops; the cloths just block up. If you are lucky enough to have a warming cabinet like mine it is possible to put the wax blocks into a conical strainer, or a large funnel with a wad of cloth in the neck. Have a container to catch the molten wax and put the whole thing in the cabinet at about 70°C. The resulting wax will be quite clean of debris.

There are lots of pieces of expensive equipment available to beekeepers these days, but I doubt many hobby beekeepers want to spend much money on processing wax. If you can somehow produce blocks of even poor-quality beeswax it can be exchanged for foundation or other goods at many of the beekeeping suppliers. There are always queues at the BBKA Spring Convention, and it doesn't matter how clean the wax is, providing it is reasonably presentable.

I'm sure many new beekeepers want to make use of their beeswax – I did. What better place to mess about with wax than the kitchen. Everything you need is there – a cooker to heat the wax, and soap and water to wash things off. Before you go down that road, stop and think. By now you will already have learnt that when extracting honey in the kitchen you can get honey everywhere, propolis will leave marks on work surfaces and floor. Beeswax is worse by far! If you should accidentally spill any hot wax it will splash over everything. It is almost impossible to remove from carpets.

I have made hundreds of candles the first of which were rather unattractive, but I did improve once I learnt, first, not to darken the wax by overheating and, second, how to make wax lighter by bleaching. The processes of bleaching, using hydrogen peroxide, was rather hit and miss to begin with, as I had no information on how to do it. I could find no guidance at all in any of the books, only a passing comment about using hydrogen peroxide. Thank goodness I had a small cooker in my shed/workshop, because the fumes from the bleaching would not have been good in the kitchen. The process worked, and it was possible to turn dark brown wax into pale creamy candles, although it does destroy most of the smell of the wax. One word of warning though – there is a small residual of peroxide in the wax, so it shouldn't be used for polish, cosmetics or making your own foundation. I once used some of my bleached wax for making polish, and it quickly made the tins rust.

See Appendix 7 for an outline of Wax Processing.

There is an art to making candles using silicone rubber moulds. If the wax is too hot it can run out of the bottom of the mould where the wick goes through the end and it will also shrink at the open end of the mould. Too cool and the wax won't get into all the little crevices in some moulds and the candle will have a rounded end to it. When it does go wrong, all is not lost because you just melt it down and start again.

I have tried metal moulds and glass moulds but struggled to get the candle out of the mould. It is probably something to do with beeswax being slightly sticky. I'm afraid I just gave up rather than experiment any more with those types of mould – silicone moulds are better.

Most beekeepers who belong to associations want to take part in the annual honey show. The classes aren't just for honey though. If you want to win a class in the show, then enter those classes that have the fewest entries, and these are often beeswax classes.

Over many years of trying I won almost every class in my local association show except for three – the runny honey classes (where I never did better than second), photography and honey cake – I'm afraid my skills, such as they are, don't extend to baking.

There is a certain satisfaction though in producing a prize-winning cake of beeswax, but you need to start at least a month before the show and there will be lots of attempts before you get it right.

9.10
Maybe not a perfect block of wax, but good enough to win a prize in your local show.

10

Drones

Drones play a vital role in the honeybee colony, but their importance is often either misunderstood or ignored. Back in the last century we didn't appear to have issues with queens getting mated, but now this is a common problem. I wonder if the trouble with queen mating is due to the quality of the drones. While I have no evidence, I believe we have lost sight of the importance of the health and wellbeing of drones. I have recently seen suggestions that *Nosema ceranae* together with viruses can have an impact on drone virility.

Many of the same factors involved in swarming control the timing and extent of drone production by colonies. Drones are produced and maintained only when colonies can support them, and when queens are potentially available for mating. The mating system has evolved so that queens can mate with many drones, most often with drones from other nests. The queen mates in the air with up to 15 or so drones. To be successful the drones have to be fit, so that they can find the queen, out-fly the other drones to catch her and then have sufficient healthy sperm to mate with her. Sadly, for them they die after mating. If the drones are not up to scratch physically, and present in sufficient numbers, then mating will be only partially successful.

Drones are often seen by beekeepers as a drain on the resources of the colony. They don't forage, they don't do any of the necessary jobs in the hive like wax making, nursing brood or defending the colony. In fact, they even drift between hives, so have been blamed for the spread of some diseases. They need feeding and so are considered by many to reduce the potential of the honey crop. One thing that I have learnt in my time as a beekeeper is not to assume you know better than the bees. They will raise drones in the quantity they require if allowed to do so. I can prove that we don't allow the bees to have enough drone comb, and you can prove this to yourself very simply. Just replace one of the brood combs with a comb from the super and see what the bees do. They will build comb along the bottom of the frame and it will be drone size cells! Why else would they do that other than they don't have enough drone comb?

When we provide foundation in the brood box, normally we put in 100% worker size. That in itself limits the ability for the colony to produce drones. I would go so far as to suggest that many beekeepers do as I did in the past and choose to replace brood combs that have a high proportion of drone cells. It would appear that the average amount of drone comb in a wild colony is about 15%, which is the equivalent of more than 1 frame in 11 in the hive. The proportion of drone comb can be as high as 35%, but this would be unusual. How many of us have the equivalent of one brood comb of drone size in each brood box? I guess that not one of us plans to normally have that quantity of drone cells, as I've never seen it stated in any books aimed at those starting out in the hobby. For most beekeepers, the only situation where they might normally increase the amount of drone comb is for mass production of drones in a queen rearing programme.

Drones are normally expelled from the hive at the start of autumn, and it is highly unusual to see any drones in the hive during winter.

One unexpected thing to me about drone behaviour is that they drift between hives and are accepted by colonies quite readily. I remember talking to a beekeeper once who had marked a large number of drones in each of the hives in his apiary, using a different marker colour for each hive. A few days later all the colonies in the apiary contained drones with all colours.

It is interesting that when we have a hive containing a virgin queen there is a very high proportion of drones present, yet it is not drones related to that queen whom she expects to mate with. Once the queen is mated and brood production is underway the drone numbers drop back to more normal levels. I once put a queen excluder under the brood box when hiving a cast swarm. When I removed it two days later, the space underneath the excluder was full of drones! It may be that drones are attracted to hives that contain virgin queens.

We have learnt that the Varroa mite is far more attracted to drone brood than worker brood, and have taken advantage of this in our Varroa control strategy. As the development period for a drone is 24 days compared with 21 days for workers, the population of Varroa will grow significantly faster the more drone cells are available. For years now many beekeepers have been using 'sacrificial drone brood' as a way of restricting Varroa. These beekeeping methods limit the number of Varroa mites, but also limit the production of drones. Unfortunately, the scenario gets worse because having reduced the quantity of drone brood cells those few drone brood cells left will be highly attractive to the Varroa mite. The few drones that do finally emerge are useless if they developed with Varroa mites in the cells.

It is almost a perfect storm: restrict the colony from raising the required number of drones and then have those few left weakened by Varroa. Is it any wonder that we don't have enough fit and active drones to mate with our queens?

Unfortunately, we are faced with a dilemma; in an attempt to reduce the number of Varroa mites by restriction of drone numbers we have insufficient numbers of fit and healthy drones. There is no simple answer. I can understand why commercial beekeepers buy 'foreign' queens in such numbers as they do and then restrict the number of drones, but that method does not conform to my beekeeping philosophy of supporting local bees.

"It is almost a perfect storm: restrict the colony from raising the required number of drones and then have those few left weakened by Varroa. Is it any wonder that we don't have enough fit and active drones to mate with our queens?"

If we are not careful when raising more drones, we will also raise more Varroa mites and that in turn results in a higher virus load on the colony.

I've often noticed drone cells in the first super above the queen excluder containing eggs. I can only assume that this is due to the pressure to produce drones when there is so little drone comb available in the brood box. I'm not sure how the eggs get into the cells though. Are they from laying workers, or do the workers transfer eggs to these cells as the queen certainly can't get through the queen excluder? I'm afraid that this question will go unanswered as I frankly don't have a clue how it happens. One consequence, however, is that the queen excluder also becomes a drone excluder, and any drones above it are trapped and can even partially block the queen excluder.

11

Feeding

I suppose from a non-beekeeper's viewpoint it is difficult to understand why bees should need feeding; after all, they have survived perfectly well for thousands of years before beekeepers came along. The difference, of course, is that we plunder the stores that they have worked all summer to produce. In simple terms therefore, if we take away their honey, we need to replace it with a cheaper substitute – sugar. I have heard non-bee-keepers argue that feeding with sugar is wrong and will dilute the honey. I counter this in two ways.

First, with our climate, even in an average year, there is often not enough honey stored in the hive for the bees to overwinter. At the end

11.1
'Are you hungry? It looks like our beekeeper has stolen our honey again!'

of summer, the brood box is only partially full and together with a full super would just about be enough for the bees to overwinter. Colonies that have swarmed, or small swarms, will often not have sufficient stores to last through winter.

Second, if we feed after taking off the honey and ensure that the colonies go into the winter with a brood box full of sugar syrup, then those stores will be used by the time the spring comes, so that none of the sugar gets into the honey.

The important thing is to feed using refined white sugar. Bees need winter food that contains as few impurities as possible as these waste products will build up in the gut. If due to cold weather the bees are unable to fly to defecate then they will end up doing so in the hive, which could lead to dysentery. For this reason, some beekeepers think that bees don't overwinter well on heather honey, as it has a high protein content. From personal experience, I have never found a problem with heather honey.

I've listened to beekeepers discuss the relative merits of cane sugar or sugar from beet. My only consideration has been which is cheapest! This particular debate may become more heated in the future, if sugar beet seed is treated with neonicotinoids. There are ready made syrups available, together with fondants and candies. There are pollen substitute feeds and various additives that apparently help health. To the beginner this is another beekeeping minefield.

It is very easy to get the main feed for winter wrong, and I'm sure that in those early learning years of my beekeeping I made mistakes.

Types of feeders

My first feeder was a half-gallon (1.9 l) contact feeder. The instructions were quite simple. 'Fill with sugar to about an inch from the top and add warm water. Stir until the sugar is fully dissolved. Continue adding water until completely full. Press the lid on firmly and quickly invert the feeder and place over the feed hole on the crown board of the hive.' Simple enough one might think. The filling and mixing part is fine, but as for inverting it and placing over the feed hole! No one told me that when you invert it a lot of liquid comes out through the mesh in the lid, until a partial vacuum has been formed. If you just invert it over the hive a quantity of sugar solution runs straight through the hive and out of the entrance. The whole operation takes some time so if you take the roof off the hive ready to put the feeder on, by the time you actually get the feeder on, many bees have run out of the feeder hole and you end up squashing them with the feeder.

11.2
Contact feeder.

This feeder was only half a gallon in volume and I thought I would need about 2 gallons (7.57 l) to complete the feed for winter, so the process is repeated three more times. The spilt sugar solution attracted bees and wasps, so with hindsight there is a good chance that robbing took place. At least I knew that you had to do it in the evening, but that just means you are fumbling around in the semi-dark making it a complete farce. You need three hands, two to hold the feeder and one to hold the torch! I could have held the torch in my mouth if it wasn't for the veil! There had to be a better way. I later learnt that the easy way to limit spillage is to invert the feeder over a bucket, wait until the dripping finished and then put onto the hive. I don't recall ever reading that bit of advice! Ensure that the lid is fully pressed on and there are no cracks in either the feeder or its rim as the feeder will just empty very quickly.

In my pile of old beekeeping equipment, which came with my first hive, was an Ashforth feeder. This feeder had a 2 gallon (7.57 l) capacity so this would surely solve most of the problems. To use this type of feeder you also need a spirit level, as it is necessary for there to be a slight

11.3
Ashforth
feeder.

tilt so that all the syrup can be taken by the bees. Just one more bit of kit to keep in the shed.

This type of feeder might solve some of the problems, but it also introduced other unforeseen ones. The feeder can be put on the hive in advance ready for the syrup to be poured in later in the evening. No more spillage or syrup dripping through the hive. First problem was the feeder was old and the joints no longer watertight, so syrup seeped through some of the joints and dripped down the sides of the hive – more bees and wasp attracted. When I lifted the roof of the old WBC hive a few days later to check how much syrup was left I found it had all gone, but the feeder now had lots of dead, drowned bees that had come up around the outside of the inner boxes. With a WBC hive this type of feeder needs a cover to stop the bees getting into the top, or with a National hive the roof must be bee proof.

I wonder how my bees survived my foolishness, but they did.

The other problem with this type of feeder is that unless the colony is quite strong they don't easily find the entrance to the feeder as it is on the edge, so you need to dribble a little bit of syrup down the access slot to help them find the way in. A similar type of feeder is the Miller feeder, which has the access slot across the middle.

11.4
Rapid feeder.

I tried *rapid* feeders. These are round feeders, that hold about 3 or 4 pints (1.42–1.89 l) of syrup. They have a cover so bees can't gain access from the top and the feeders can easily be refilled without removing them from the hive. Great except that you have to refill multiple times to give the hive enough for the winter.

I recalled that from those early school days feeders were made from 2 gallon (7.57 l) honey tins with a series of small holes made in the lid, much like the modern contact feeder. Alternatively, 15 kg honey buckets are cheaper than buying feeders and if you punch some holes in the lid using a nail these can be used as a contact feeder. A 15 kg bucket contains about 2 gallons (7.57 l) so you can do all your winter feed in one trip. They are large enough so that the roof of the hive will balance on top without needing another empty box. All OK until the feeder is empty and a bit of wind will blow the roof and feeder off. Note to self at the time – *put a hive strap over the whole thing to keep it together.*

More recently at the BBKA Spring Convention I came across the English feeder. This is a large rectangular rapid feeder, which contains about 6 litres or just over 1 gallon. It is shallow enough that a standard roof fits over. It has a central feed hole which goes over the central hole

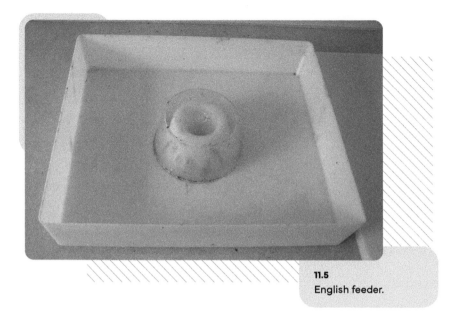

11.5
English feeder.

in the crown board and a lid to make it bee proof. I've used this type of feeder for many years and it works for me.

When to feed and why

I think I've already answered the question about when to feed, but not why we should feed in the evening. Once the bees have settled down for the day and have stopped flying, feeding can be done. When bees find a food source, they use dances to indicate to others where a food source is. Beekeeping books describe the waggle dance for food sources at a distance and the round dance for a source close by. I've not seen the dance they do for 'You might not believe this, but Geoff has put some food in the attic!', but when they find food in the feeder, they will get the message that food is nearby. If you feed in the evening, by the time the bees get going in the morning they will have found the feeder and won't go out looking for food. Once this first feed has been done you can refill the feeder at any time of the day without causing any disturbance. Hopefully, during all this process you won't have spilt any syrup. If you have then you will have the bees in frenzy everywhere, once the bees learn that there is a food source close by.

I once tripped over a bucket of syrup that I had put down while pouring feed into another hive. There was syrup everywhere, including on me, and the following morning there was a feeding frenzy in the apiary. (I put this down as an accident rather than a mistake, although I should have been taking more care.)

Beware of carrying buckets with plastic handles as I've known them to break. I now transfer the syrup into plastic jerry cans as these are easier to carry and less liable to spillage than buckets. I used to take food to my out apiaries using my pickup. The 15 kg honey buckets with lids on were put in the back. One day I had to brake heavily, and the buckets all slid forward. Some of the lids popped off and I found I had 2 l of sugar syrup in the back of the pickup. The jerry cans are much safer.

While I might have implied that the problem of attracting bees and wasps during feeding is just a bit of a nuisance, it does really need to be considered seriously, as robbing can be difficult to eliminate once it has started. To enable a hive to successfully defend itself from robbing, it is essential that feeding is done only once the hive entrances have been restricted to quite a small opening. A robbed hive will soon die of starvation.

The idea is to get main winter feeding done once the honey has been taken off and either before or after the main Varroa treatment. Some treatments and feeding can be done simultaneously, but many cannot. If your preferred method of treatment is going to take 28 days and feeding is not possible during that time, then your timetable becomes quite tricky, and the bees might starve during this time if you have just taken off all the honey! Feeding too early will just stimulate more brood production, whereas what we want the bees to do is to fill the empty portion of the brood box with stores. As the autumn evenings get cooler the bees can't process and seal the stores. Unsealed stores will pick up water and become diluted and then ferment. Fermented stores will cause dysentery, so feeding too late in the year will cause problems for the bees. As a rule of thumb, I like to complete feeding by the end of September or early October at the absolute latest.

The autumn feed should be with a concentrated sugar syrup made up of two parts white sugar to one part water (hot water will help to dissolve the sugar), or you can buy a readymade specialise bee feed syrup. The quantity of feed really depends on the type of bee you have. The books normally recommend about 18–22 kg of stores needed for the winter. If you have Italian bees then it will be considerably more, but if like me you have a near native strain of bee they will overwinter on as little as 10 kg. A National brood frame contains about 2.2 kg, so 11 frames will be sufficient for most bees. If you have more prolific bees that have large winter populations then you will not only need more stores but will require a larger brood box.

Feeding is quite a common topic of conversation at association meetings in September or October, when beekeepers often discuss whether all the autumn feeding has been done. I once heard a beekeeper in November who was still feeding and he said the bees were still taking the syrup down. Well, so they might, but were they able to process it and seal it? I never did ask that beekeeper if the bees had dysentery in the spring. Sometimes we can kill them with kindness!

Many beekeepers, who don't want to maximise their honey crop, like to leave supers of honey on the hive for the winter rather than feed sugar syrup. There is no problem with the theory as bees have of course overwintered better on their stored honey for thousands of years. The problem with leaving a super on comes with the way the bees use the stores over the winter. The winter cluster tries to stay in contact with the stored food. As the stores are consumed the cluster moves around the hive, progressing upward as the stores are used. If the last of the food is in a super, separated from the brood box by a queen excluder, then not all the bees in the cluster will get through the queen excluder. Actually, all but one will be able to get through. Should the cluster actually progress right through the excluder so that none of it remains in the brood box, unfortunately the queen will be left behind with just a few attendants and no food. In the spring you will find a cluster of bees in the super and a small dead cluster, with queen, in the brood box.

The lesson if you want to leave honey on for the bees is to remove the queen excluder. In the spring make certain the queen is back in the brood box and then put the queen excluder back on.

If you have drone comb on the supers then there will probably be drone brood in the super, so when you put the queen excluder back on there may be drone brood above the excluder, which will also keep the adult drones out of the brood box and their access to the outside.

There are other types of feeder for different situations. Rapid feeders, including Miller or Asforth feeders are great for autumn feeding, but are of no use if you need to feed in the early part of the season, as the bees won't leave the winter cluster to go up into the feeder. They will take food from a contact feeder, but only if the cluster is right underneath the feeder. If the bees were fed correctly in September, they won't need feeding in the early spring. It seems to be a constant worry of beekeepers in general through the winter that the bees are going to starve. Unless you have some of those hungry foreigners, that need far larger winter stores, they seldom need any additional food. There are various types of fondant or candy available, some with added pollen substitute or additive, which is designed to assist digestion or are sold as effective in combating Nosema, European or American foul brood. I can't comment on these as I've never tried them or felt the need for them.

Fondant is useful as an emergency food if the colony is short on stores in January or February and I did use them in the past when I was less confident in my beekeeping. Fondant can also be used as a spring stimulant to get the colony going early so that they are nice and strong to take advantage of the oil seed rape. I don't have any oil seed rape within flying distance in our North Wales hills, so I don't bother with a spring stimulant.

Liquid feed can't be fed again until March, but if they do need it then the strength should be one part sugar to one part water and, if possible, fed with a contact feeder.

"Fondant is useful as an emergency food if the colony is short on stores in January or February"

11.6
Plastic frame feeder.

There are other types of feeder for special situations, the most common of which is the frame feeder. This type of feeder takes the place of a brood frame on the edge of the brood box but is most likely to be used when feeding a nuc during the active season. They used to be made from wood and as they got old they would leak.

New ones now are made from plastic, so they are fully waterproof. Depending on the manufacturer, however, they do have an issue with the float which goes on top of the reservoir of syrup. The ones I've used had a plastic float that was too small and bees drowned in the syrup. A small handful of wood shavings solved the problem, but a bigger float would have been better.

Large feeders are useful during the active season for feeding a swarm to enable it to draw out comb rapidly, or for feeding a shook swarm or Bailey frame change. Syrup strength need only be 1 : 1 for this situation, as we want the bees to use it, not store it. Stronger syrups can be used though if that is what you have at the time.

11.7
Very small colonies or nucs can be fed by using small honey jar contact feeders.

Emergency feeding

Best practice should avoid the necessity for emergency feeding, but even in the best managed situations weather conditions can lead to potential starvation. Bees can starve at any time of the year and usually if this happens it is the beekeeper and not the bees who should take the blame. It is the responsibility of the beekeeper to make sure that bees have enough stores. Sometimes in winter bees use up stores on one side of the hive and become marooned away from stores elsewhere. This is known as isolation starvation. Frames of stores can be moved across so they are adjacent to the bee cluster. Do not divide the brood nest. Of course, you can only do this if you have opened up the hive to check, and we tend not to open hives in the middle of winter.

Many beekeepers check for stores in the hive during winter months by 'hefting', which is lifting one side of the hive to check how heavy it is. That is OK, but if all the stores are on one side the hive can actually appear heavier (or lighter) than it really is.

If bees are short of stores in the winter and likely to starve then white soft candy or bakers' fondant is placed over the crown board feed hole. In the case of small colonies, the crown board may need turning in order to position a feed hole over the bee cluster, or the crown board taken off altogether, so that the fondant can be put directly onto the top bars of the brood box. Bees require water, often taken as condensation within the hive, to make use of candy. Candy is therefore taken slowly and does not excite the colony as much as other feeds. If sugar syrup is offered in a contact feeder, changes in temperatures may cause expansion and contraction of the container, pushing syrup through the mesh and wetting of the cluster.

In extreme cases during the summer season if bees are starving spray them with a thin sugar syrup solution and fill an empty comb with sugar syrup. This can be done by pouring the syrup into the cells slowly by using a squeezy bottle, for example, a clean washing up bottle. When filled, place the comb adjacent to the bees. This situation might occur after a honey crop has been taken off, or a recently hived swarm often uses all its stores if the weather is inclement. They can then be given a full feed using a rapid or contact feeder.

At the harvest season when removing the honey crop always check that sufficient stores remain to prevent bees starving. Feed immediately if needed.

Remember, March and April are the months when the bees will be using up food reserves fast as the colony expands and produces more

"In extreme cases during the summer season if bees are starving spray them with a thin sugar syrup solution and fill an empty comb with sugar syrup."

brood. It is far better to have fed sufficient stores or have left lots of honey in the autumn than to do emergency feeding in the spring. At this time a colony should have at least 4–5 combs with honey/stores, that is, 9 kg or 20 lb.

Water

Many books mention the need to provide water, and one can buy water feeders for use in time of drought. It is notable that feral honeybee colonies tend to follow water courses when they swarm, indicating the importance of water to them. Bees will find sources of water, which they use to dilute stores in order to eat, or in extreme hot weather will use water to assist in cooling the hive. Some books recommend providing a water supply so as to avoid the situation where your bees collect water from a neighbour's property and cause a nuisance. The only instance I've had of this was a relatively new beekeeper who found that his bees were collecting water from a pile of wet sand at his neighbour's house. It is very difficult to stop once it has started as the bees will always go back to their source.

Bees have a preference for water that is warmer than 18°C and also for urine to which, like other insects, they are attracted by the salts contained. So, when first supplying water, adding a little salt might encourage the bees to use it. Their favourite water source appears to be from cow pats!

I visited some beekeepers in Cyprus in springtime a few years ago, where the climate is very different. Daytime temperatures in summer are commonly 40°C or above. Everywhere becomes parched. It was common to have a 50 gallon (190 l) drum, with a float on the water, which was refilled every few days (see overleaf).

One beekeeper I met there had had to move some of his bees when building development had got close to his apiaries, as every new house had a swimming pool and that was where his bees went to collect water.

With climate change and ever increasing hotter and drier summers, perhaps we will need to pay more attention to provision of water for our bees.

11.8
Large water feeder in an apiary in Cyprus.

11.9
Polystyrene float in the large water feeder.

"With climate change and ever increasing hotter and drier summers perhaps we will need to pay more attention to provision of water for our bees."

12

Management through the season

How we manage our bees is down to individual preference. A commercial beekeeper with hundreds of hives will have a totally different regime to a small-scale hobbyist beekeeper with just a few hives.

In a large-scale undertaking, the beekeeper does not have time for weekly inspections for swarm control, so will employ a system designed to restrict swarming. The criteria used are to have young queens and to ensure the bees have sufficient space for brood and for honey storage. It is this requirement for young queens that is the reason for so much conflict between various factions in beekeeping. Many commercial beekeepers have imported large numbers of package bees in the spring, mostly from Italy. Alternatively, they import queens from parts of Europe such as Italy, Greece or Cyprus that can supply queens early in the season. Either way, package bees or colonies requeened early in the season are unlikely to swarm provided the bees have sufficient space to expand.

From a purely commercial point of view this looks to be a perfectly reasonable way to run a business but looking at the overall position within beekeeping in the UK I believe it is far from acceptable, for the reasons outlined earlier in Chapter 8.

The small-scale hobbyist beekeeper needs to adopt a different approach to management through the season. There will be a different emphasis depending on whether the beekeeper wants to maximise honey production, or merely to enjoy the hobby of beekeeping.

If you want to maximise honey production, then first you need to understand what plants locally will provide a crop and when.

In 2012 I had some of my honey analysed by our local Trading Standards Services. The honey had been extracted in late summer, but the pollen analysis described it as 'characteristic for a Welsh honey mainly from a spring harvest'. There was no 'frequent pollen'. It contained a big mix including sycamore, chestnut, sweet chestnut, dandelion and willow, with very small amounts of clover. My guess is that it was at least 75% spring honey, so what happened to the summer honey?

Unfortunately, many of the beekeeping books are out of date with regard to the sources of nectar. One summer nectar source often quoted is clover, however red clover, while excellent for bumblebees with long tongues is useless for our short-tongued honeybees. White clover has been bred to maximise nitrogen fixing in the soil, but also has flowerets that are too long for honeybees. Only the old wild white clover will produce a reasonable honey crop, if the weather is warm enough. Unfortunately, the old meadows with wild clover have long since disappeared. While there may be some clover in your honey, it is a mistake to think that you will get a honey crop from clover. I remember honey from my childhood that was clover honey, softly granulated with large crystals. Unfortunately, those days are gone.

Large arable areas of the UK have huge acreages of oil seed rape (OSR), mostly autumn sown and this will flower in April. Spring sown varieties flower around June. While these crops are self-fertile, they do produce huge quantities of nectar, and beekeepers in areas of OSR will have to adopt a different approach to management than elsewhere. There is no OSR grown within about 8 miles (13 km) of my home, so I know my bees do not have the opportunity to forage it. I regard this as a bit of a blessing as my honey is a mixed floral spring mix. These floral sources provide a honey high in fructose and low in glucose, so granulation is slow. With OSR the glucose content is high and granulation is very rapid. I also find the taste and aroma of OSR honey rather disappointing but mixed with a stronger flavoured honey it makes a good soft set blend.

If you have OSR on your doorstep then you need to be prepared to handle the situation appropriately. The season will start early and may catch you by surprise. If you want to maximise the crop then you need to stimulate the bees to produce brood early in the season, before the honey flow starts. You can do this by feeding them with fondant and/or pollen substitute in late February and early March. I have to admit that our native honeybee is slow to build up in the spring and I can understand why other beekeepers might prefer other strains of bee.

Once the honey flow from the OSR starts, activity in the hive becomes markedly more intense. Irrespective of how much brood there is the bees will bring nectar in and use every available space. We need to pre-empt the situation by providing more space ahead of time.

As OSR honey granulates very rapidly, many beekeepers find it has granulated in the combs before it can be extracted. The combs then have to be melted down on a heated uncapping tray. Knowing that this is likely to happen, the frames in the supers will have foundation, rather than being already drawn comb. This presents us with a couple of problems. First, early in the season bees don't like going through a queen excluder (especially if it is a plain slotted one – see the pictures in Chapter 2) and they don't readily take to leaving the brood box to go onto foundation.

The mistake that is often made is to put on a queen excluder followed by a super containing foundation. The bees reluctant to go through the excluder onto the uninviting foundation become overcrowded in the brood box and prepare to swarm. A simple modification to this management practice should overcome the problem. First, don't put on an excluder. Put the super directly on top of the brood box. After a few days the bees will have ventured into the super and started to draw out comb. At this stage put a queen excluder between the brood box and the super, making sure that the queen has not been trapped above the excluder. The simple rule of thumb about when to put on a super is when the topmost box on the hive has bees on 75% of the frames. This doesn't always work with rapid nectar flows as you get with OSR. It pays to keep well ahead of the bees and if in doubt put extra supers on. The worst that can happen is that the bees will draw out and fill only the central combs, but you can always rearrange the frames later if you want to.

With a spring flow from sycamore and dandelion, things don't tend to be quite as hectic, but the same principles apply if you only have supers fitted with foundation. With drawn combs, particularly when they have been stored wet, these precautions are not needed. Putting an excluder on and then a super of wet combs will result in the bees moving up within minutes.

Springtime should be the best time to think about moving the bees onto fresh brood combs, but if the spring flow is all you are going to get then perhaps it is better to wait until later in the season.

Once the trees and dandelions have finished in my region it is almost like a green desert, and it will be no different once OSR has finished in arable regions. Every year I have the dilemma of either taking the spring crop off and potentially having to feed the colonies, or leaving it on and taking the whole crop off at the end of summer. I generally leave it on and then later on wish that I'd taken it off, as often the bees will eat much of the spring crop as the season goes on, particularly if the weather is poor. With OSR you will have to take it off and extract immediately, and possibly have to feed the bees as well.

I am finding now that the input of honey during summer can be limited, even assuming the weather is suitable for the bees. There used to be a June gap in forage, but now this is not so marked – forage has just generally reduced in late May, June and July. Half a century of grassland 'improvement', the lack of field margins and mechanical hedgerow management have almost totally removed summer wildflowers!

I had an out apiary on a field that was set aside and as an experiment I agreed with the farmer to sow about an acre of phacelia.

This crop is normally grown as a green manure, but we let ours flower. The field was alive with bees of every sort. The honeybees didn't produce much noticeable honey from it, but the pollen they collected was bright blue.

12.1
Phacelia.

As there is no honey input in these early summer months, surely that is the time to do our frame changes and later in the summer to do our main Varroa treatments. Unfortunately, all our outdated beekeeping books tell us to do things differently to that.

When I took bees to a field of borage two decades ago, they each filled six supers in about a month. Failure to ensure they have enough space would guarantee that they will swarm.

12.2
You may be lucky enough to be near a field of borage, if so be prepared.

12.3
Hives on the heather moor.

Late in the summer there are two big sources of honey, and these both pose problems with timing of Varroa treatments and subsequent winter preparations.

The first of these is heather. I'm blessed that I live about half a mile from the edge of the heather moors on the Clwydian Hills, so weather permitting, my bees will forage on the heather, from late August through to early October.

12.4
Bees on heather with pollen all over their body and legs.

Generally, as the weather can be so fickle, the honey flow is restricted to about 2 weeks during that period. Yields from the heather can be quite variable, often depending on how the heather is managed. Old neglected plants yield little and are also not grazed by moorland sheep. The plants need to be young for both nectar production and grazing, so good management by burning or flailing is essential.

12.5
Himalayan Balsam. Look carefully and you can see a bee inside the flower.

The other potential late summer crop is from Himalayan balsam. This invasive species is causing big environmental issues throughout the country.

Primarily it is found on the banks of rivers where it out competes the other vegetation, so that when it dies back the banks are left bare and then get eroded. The seedpods explode sending the seeds high into the air and are transported by water downstream. Environmentalists are working hard to try to get rid of the plants, however, from the beekeeper's standpoint this is a valuable source of nectar. The plants flower from late summer onwards and continue well into autumn. The stamen of the flower at located in the top part of the flower trumpet and the bees pick up the pollen on their backs, where they find it very difficult to remove. The pollen is white in colour and the backs of bees returning from foraging trips are covered in white. They look like ghost bees.

The honey is a valuable bonus for the beekeeper, but it does interfere with the timing of Varroa treatment. It might be a mistake to try to administer treatment in September if the bees are bringing in nectar from balsam.

The lessons to learn therefore are to not rely too much on what the older books state about nectar sources. Watch what the bees are bringing in and the quantities involved. The colour of pollen loads will give you an indication of the possible nectar sources. Any nectar flow will be weather dependant, so will vary from year to year. Once you have established a pattern then you can vary the timing of honey extraction and Varroa treatments accordingly. Do not rely on some of the book recommendations and timings, as they are often based on average weather conditions, average bees and average availability of forage and are decades out of date. No two areas of the country are the same, and no two apiaries are the same. There is also variation between hives in the same apiary.

13

Making increase and uniting

As bees live in a colony with just one queen, in order to increase, or even just to maintain their numbers, they have evolved the swarming mechanism where approximately half the colony leave with the queen and leave a new queen or queens to emerge and allow the old colony to rebuild; this is called swarming. In theory they could do this every year, so there could soon be a situation where we would be overrun by bees. Clearly this doesn't happen as there is a large mortality rate in wild colonies. Many swarms in the wild don't live long enough to make it through the following winter.

A simple way for beekeepers to make increase is to just allow the colony to swarm and then catch the swarm and put it into a new hive. In ancient times when bees were kept in skep hives (domed baskets) this would be precisely what was done. This is a rather hit and miss system.

In the modern era with swarm control techniques it is possible to intervene and to do some sort of split and thus increase the number of hives. Even allowing for significant winter losses it is still possible to maintain the number of hives, or to increase in number. In fact, it

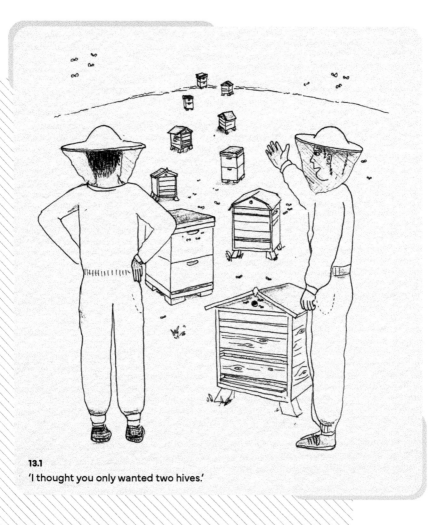

13.1
'I thought you only wanted two hives.'

is far easier to increase the number of hives rather than decrease. For many years I have wanted to reduce the number of hives that I had, and although I sold some I still ended up with just as many at the end of the season as at the start.

I remember about the turn of the century there was a large interest in beekeeping courses, which resulted in a big demand from the new beekeepers for nucs. What I learnt then was that it was possible to make as much money from a hive by selling bees as it was from selling honey.

Unfortunately, some commercial suppliers already knew this and were supplying starter colonies with imported queens of dubious parentage. Although it took a couple of years to organise, like many other local associations, we managed with a little encouragement and minimal instruction to get our members to produce sufficient surplus of our local bees to meet the demand. Some beekeepers are reluctant to make up nucs for sale, as it is part of beekeeping folk law that it is bad luck to sell bees. It seems to me to be worse luck to have a new beekeeper set up nearby with bees of a different race that may well be the underlying cause of bad-tempered bees!

Sometimes circumstances are just right, other times the weather or other things outside our control thwarts us.

One summer I had eight virgin queens that I had raised that I was going to put into mini nucs for mating. Two large colonies also needed to have artificial swarms done. I did the artificial swarms and took the queenless portions of the hives away to an apiary a few miles away. The following day these hives were split into eight five frame nucs, using some extra frames to make up to the required number, and a virgin queen introduced into each. Conditions must have been ideal because each queen successfully mated and the new colonies quickly expanded.

See Appendix 8 for methods of making increase.

One mistake that some beekeepers make is to expect a nuc sized colony to raise a good queen. To get a good queen, the larva has to be well fed from the beginning of its life, and that can only be done in a large colony that has plenty of nurse bees. It is possible to make up a small queenless nuc and provided there are some eggs they will try to raise a queen by making emergency queen cells. The resulting queen will often be disappointingly small and will be short lived. A queen raised under emergency conditions in a large healthy colony will be as good as a queen raised by other methods. A colony which has produced multiple queen cells can be used to make up nucs, each with a good-sized queen cell and in this instance the queens will be as good as any. The colony does not have to be large to support the queen until she has mated and started to lay eggs, but it does need to be large to actually raise the queen cells.

For some the difficulty is not making an increase, rather how not to make increase. Having learnt to do some form of split for swarm control they never learnt how to combine two colonies. I met one beekeeper who had about 12 hives who said that, really, they never wanted that many, but every time they did their swarm control they ended up with another colony.

The important thing to remember is that the bees cannot just be put together as they will fight. The uniting process needs to be slow so that the bees acquire the same colony smell. Even re-uniting an artificial swarm with the parent hive once the new queen has been mated requires care. First, the two colonies to be united have to be moved as close together as possible. I've heard of beekeepers forgetting this vital step and find all the flying bees try to go back to the old site in the apiary, where they will eventually all drift into the next nearest hives. If possible, do the uniting in the evening when most of the flying has ceased, disturbing the bees as little as possible. If you have the time it is best to prepare during the day, find the queens and decide which one you are going to keep, and then come back later to actually do the uniting. As with all operations with bees – plan ahead and get all the equipment needed together. The hives will be full of bees, and that is why you prepared everything earlier and found the queens. It would be impossible to find the queens in such full colonies once the evening light starts to diminish.

In ideal circumstances that is what you should do, but in my experience, when you have lots of hives and are on a tight schedule you just have to do it at whatever time of day it is.

"The important thing to remember is that the bees cannot just be put together as they will fight. The uniting process needs to be slow so that the bees acquire the same colony smell."

The first scenario for uniting is when both colonies to be united are the same size. Remove all the supers and put them to one side. Decide which queen is going to be kept and remove the other (you may want to keep her for use elsewhere, in which case put her in a travelling cage with a few attendants). If you want to be extra cautious, then put the queen you are keeping in a cage to keep her safe while the colonies get united. Put a sheet of newspaper on top of the brood box of the colony with the queen (whether caged or not) and then put a queen excluder on top of the paper. The excluder is not essential, but it does keep the newspaper in place – vital if there is any wind! Make two or three small holes in the newspaper with the corner of your hive tool and put the other (now queenless) brood box on top. Broadsheet newspapers work better than tabloid as a single sheet will cover the entire hive. Some beekeepers advocate making the holes with a pin rather than the hive tool, as the idea is to allow the bees to get to know each other slowly and thus avoid fighting. I've always used the hive tool and it seems to work. If you have caged the queen, you will need to go back into the hive after a couple of days to let the queen out.

We now have the problem of what to do with the supers as they are full of bees from both hives. Put the queen excluder back on top of the second brood box followed by the supers from that hive. I've been told that the bees in the supers won't fight, but I prefer not to take any chances. I put another sheet of newspaper on top of the supers followed by a queen excluder. Again, make some holes in the paper and put the remaining supers on the top, if you can reach of course as by now it could be quite high! If the hive is really too high to reach at this stage, then I suggest the extra supers could be put on another hive in the apiary. If the whole operation had been planned in advance, then the bees could have been cleared down from the excess supers to avoid the problem of the united hive being too high. Finally put the crown board and roof on. Leave for a week and then open the hive up and remove the queen excluder from between the two brood boxes.

Some beekeepers when uniting two hives, both with supers, will keep the bottom hive and supers together and then put the newspaper plus queen excluder on top of the super. The other brood box and supers then go on top of the pile. You then go back and sort out all the boxes a week later.

If you are one of the thousands of beekeepers who can't find the queen, you just have to take a chance and do it with both queens still in the hive. If one is an old queen and one is young, the young one normally survives. In this situation some beekeepers like to use two queen excluders, with a bee space between the two, so as to ensure the queens don't try to fight through the queen excluder. Once the excluders are removed of course they will have to meet up. Unfortunately, as is so often the case the bees have not read the book and it doesn't always go quite according to plan. Strangely, the more often I united colonies the more successful it was.

If uniting a large colony with a small colony or queenless one I prefer to put the small one on the top of the larger one. When uniting two smaller colonies it may be better to have both colonies in full size brood boxes, ensuring that the frames are vertically above each other and use dummy boards to restrict the widths of the two sets of combs.

An alternative way that I use to combine two nuc size colonies is to use icing sugar. The method is to take one frame out of one of the nucs and shake icing sugar over all the frame and bees. Put it into an empty brood box and then take a frame from the other nuc and do the same, putting it into the box next to the first frame. Continue the process until all the frames have been covered in icing sugar and placed alternately in the box. Shake icing sugar over any remaining bees in the nuc boxes, and tip them in on top and close it up. By the time the bees have cleaned themselves of all the icing sugar they will all have acquired the same hive smell. You can use flour; it works just as well, but results in a nasty mess outside the hive and probably attracts vermin.

It is possible to combine three or more swarms by just shaking them all together into a hive. I haven't been fortunate enough in recent years to have that many, but I recall once in the 1980s collecting 16 swarms in 1 week and ran out of hives to put them all in. Lots of them got dumped in together and they all did well.

A word of warning on uniting, if you are not sure that a colony is healthy don't unite it! If it is small, due to some unknown health problem, and you then unite it you will have just spread the problem.

Conclusions

During all my beekeeping years, I experimented, often getting it wrong, but remember Einstein's words 'Anyone who has never made a mistake has never tried anything new'. We must try to bear in mind one thing – it is wrong to think 'practice make perfect', it doesn't, 'practice makes permanent'. If you find things don't quite work out right, you need to change rather than carrying on the same way. You may not even be aware that there is an alternative or perhaps easier way to do things. We all make mistakes along the way. For those readers who have kept bees for many years you will no doubt be able to relate to some of the mistakes that I have outlined. For those of you just starting out with bees, hope-fully my words will help you through the beekeeping minefield that awaits you.

It's good to try something new and see if it works. There are always beekeepers, including me, who will tell you that their way is right. Don't just take these on face value! There are new types of hive that come on the market, they aren't necessarily an improvement on what we have and are often just an expensive gimmick. One could work solely on the philosophy 'if it ain't broke – don't fix it', but how do we know that it is broken unless we give some of these ideas a try?

Beekeeping is a rewarding hobby, and for some a commercial busi-ness, but don't make the mistake of thinking by keeping bees we are helping to conserve bees in general. The honeybee is generally well man-aged throughout the world and is given treatments to help it survive pests and diseases. Wild bees, however, are under threat due to loss of habitat and forage. It can be argued that having too many beehives in a particular location will result in the wild bees being outcompeted for what are often thin resources. We hear of companies with large land holdings paying to put hundreds of beehives on their land as if it were an environmental benefit. The only benefit will be to the beekeeper not the other insects, which will be in competition with the honeybees.

The other environmental myth is that providing large acreages of nectar and pollen rich crops such as oil seed rape will help wild bees. It is only the honeybee that stores honey in great quantity. Whereas bumblebees only store small amounts and they need flowers throughout the summer. What farmers need to provide is not large numbers of beehives, but to have well managed hedges and field margins that will encourage diversity for insects and other small animals.

One idea you should never try, in my opinion, is to introduce an imported queen. Many, as I did, fall into the trap of thinking these queens are better. They may be good tempered and highly productive in the first year, but in subsequent years all your hives and those of your neighbours may well turn out to be highly aggressive. Do not think that Buckfast bees are British. They may well originally have been bred in Buckfast, South Devon, by Brother Adam but all Buckfast strains, to my knowledge, are now imported. My advice is to buy local and to select from what you have. If you can, try to have a near native strain of bee.

Each phase of my beekeeping had different challenges. Initially I enjoyed just having a few bees and producing enough honey for family, friends and ourselves. When I became a small commercial beekeeper, the business made money from honey production, the sale of nucs and queens, and we had a regular farmers' market stall that sold a wide range of jams and pickles containing honey. The most popular product we ever made using honey was pickled onions. One year we peeled, pickled and sold half a tonne of onions. The recipe is now a well-guarded family secret, but I still get asked if we are making any more of our pickled onions. Times change, the farmers' market movement locally now seems to have run its course.

One successful sideline was wedding favours, which even sold internationally. That part of the business dried up when others started doing similar, but dare I say inferior, favours at lower prices.

When I studied for the BBKA written examination I took two modules at each sitting. It was too much and I failed the module on queen selection and breeding. I retook that module two years later and passed with distinction. If I hadn't been in such a rush to complete all the modules, I would probably have passed them all with better marks. Hindsight is a wonderful thing.

Life moves on and we get older. Now that I'm in my 70s, I couldn't manage 60 hives of bees, or extract and bottle a tonne of honey. Beekeeping on a large scale is hard and heavy work. The biggest mistake would have been to try to carry on too long! Now is the time to just pass on my knowledge, such as it is, to others. I've lost track of how many times that it was suggested 'You should write a book.' So here it is!

APPENDICES

APPENDIX 1
ARTIFICIAL SWARM

1 Artificial swarm without finding the queen

Hive with queen cells

→

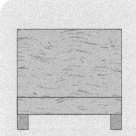

1. Move the old hive about 1 to 1.5 m to one side.
2. Set up the new floor and brood chamber with no frames on the site of the old hive.
3. Take one frame of brood from the old hive and place it in the middle of the new brood box.
4. Taking each brood frame in turn shake all the bees off the combs into the new box. NB any frames with queen cells, gently brush the bees off rather than shake!

↙

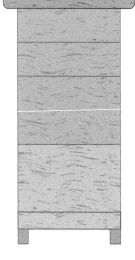

1. Put the queen excluder onto the new box and place the old brood box with the old frames on top.
2. Replace the supers and leave for about 1 hour.
3. Most of the bees will go back onto the brood frames in the top box, except for those left on the frame in the bottom box.

↗

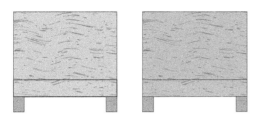

1. Open the hive and replace the old brood box on the old floor. The queen will be on the single frame in the new hive (bottom box)!
2. Remove the queen excluder and fill the new box with fresh frames of foundation.
3. Put the queen excluder on the new hive and then the supers.
4. Go through the old hive and remove all but one unsealed queen cell. Mark the frame that the cell is on.

↓

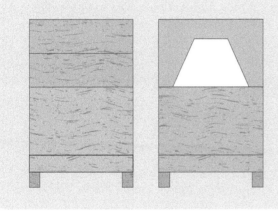

1. The old hive may need feeding. Alternatively distribute the supers between the two hives.
2. Leave for 7 days.
3. Inspect the old hive for any new queen cells and remove them. Take care not to shake or damage the queen cell previously selected.

The Hedden addition

This extra manipulation to the standard artificial swarm involves moving the old colony to the other side, on day 7. Do not move the colony with the queen. This results in additional flying bees returning to the original location.

2 Making a queenright nuc without finding the queen

Hive with queen cells →

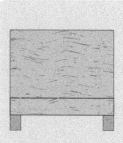

1. Move the old hive about 1 to 1.5 m to one side.
2. Set up the new floor and brood chamber with no frames.
3. Take three frames of brood (with no queen cells) and two combs of stores from the old hive and place it in the middle of the new brood box.
4. Taking each brood frame in turn from the old box, shake all the bees off the combs into the new box. NB any frames with queen cells, gently brush the bees off rather than shake!

↙

1. Put the queen excluder onto the new box and place the old brood box with the old frames on top.
2. Replace the supers and leave for about 1 hour.
3. Most of the bees will go back onto the brood frames in the top box, except for those left on the frames in the bottom box.

↗

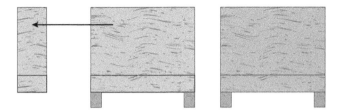

1. Open the hive and replace the old brood box on the old floor. The queen will be on one of the frames in the new hive (bottom box)!
2. Remove the queen excluder and transfer the frames into a nuc box together with two combs of stores. Shake any remaining bees into the nuc box.
3. If the nuc is to remain in the apiary shake some more bees from two other brood frames into the nuc box, as some bees will fly back to the old hive.
4. Put the old hive back in its original position and remove all but one unsealed queen cell. Mark the frame that the cell is on.

↓

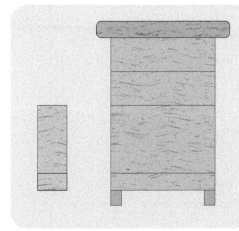

1. Put the old hive and supers back together.
2. Leave for 7 days.
3. Inspect the old hive for any new queen cells and remove them. Take care not to shake or damage the queen cell previously selected.

APPENDIX 2
MOVING BEES

When moving bees a distance I have found the following method works well.

1. Don't inspect the hive immediately prior to preparing for a move as it will break the propolis seal that the bees have produced. This will help to stop the frames moving during the move. Ensure that hives have sufficient stores as they might not be able to forage at the new location.

2. Prepare the hive for moving during the day, putting on a travelling screen and strapping the hive with two straps parallel to each other, not crossed over! Replace the roof.

3. In the evening when the bees have stopped flying, seal up the entrance with a piece of foam rubber and remove the roof.

4. If the move will only take a few minutes and the destination has been prepared beforehand then the bees can be loaded onto the trailer/pickup or car. If the journey will take some time and the new site needs preparing, then it is better to leave loading up until the morning of the next day.

5. Try to complete the journey before the temperatures get too high, particularly if the bees are being transported inside a vehicle.

6. For convenience sit the hive inside the upturned roof, as this will save space.

7. Some books recommend placing the hives with the frames pointing in the direction of travel as this limits the movement of the frames when accelerating and braking. In reality if driving carefully this requirement is unnecessary.

8. Travel to the destination without stopping if possible, but if the journey is long and a break is required then the bees might need some water pouring into the top of the hive to aid cooling.

9. Some beekeepers advocate not turning the engine off while stopping as the change in vibrations may affect the bees.

10. On arrival at the destination, set up the hives on the hive stands.

11. It is often recommended to remove the entrance foam before replacing the roof as the bees panic when the roof is put on. I've never found any problem with putting the roof on prior to removing the entrance foam.

12. Don't leave it too long before opening the hives! As soon as all the hives are set up remove the foam and put on the roof on each hive. Start with the hives furthest from your vehicle so that you don't have to walk past a hive that has already been opened.

13. Sometimes the bees will start coming out as soon as the foam is removed, other times they will not make any effort to come out.

14. Leave the hives strapped up until the next visit. Some beekeepers like to unstrap the hives before opening the entrance, as leaving the straps on makes it easier for rustlers to take the hives if the bees are to be left in an open situation such as the heather moors.

With all hives now having open mesh floors, for short moves I dispense with the use of travelling screens, as the bees have ample ventilation through the floor. The whole hive including the roof can be strapped together. It is necessary to ensure that there is ample ventilation under the hive, so the hive should sit on a hard surface. The carpeted back of the car might not be ideal, so a ventilated screen could be better, but on a trailer or back of a pickup they are fine.

"Some beekeepers advocate not turning the engine off while stopping as the change in vibrations may affect the bees."

APPENDIX 3
RECORD CARDS

RECORD CARD   Year _____

Hive No _____   Queen Marked _____

Date	Queen Present	Queen Seen	Frames Brood	Temper	Weather	Stores	Comments

Notes on record card

On 'Queen marked' record colour and date marked.

Record temper on a scale 1–10 where 1 is bad and 10 is excellent.

Stores – Q. are there enough stores until next inspection?

Record Q cells, Varroa, abnormalities and assessment of space in comments column.

APPENDIX 4
WAX MOTHS

There are currently two moths that present problems to beekeepers in the UK. They are the lesser wax moth, *Achroia grisella*, and the greater wax moth, *Galleria mellonella*. Both wax moth species undergo complete metamorphosis and have four stages of development: egg, larva, pupa and adult. Generally speaking, wax moths are considered a minor pest of the honeybee, however, they can cause problems for weak hives.

What are the problems that wax moth cause?
Wax moth and the larvae do not kill honeybees or their colonies but are often found destroying unoccupied drawn comb in colonies that have failing queens, been damaged by pesticides, are weak, queenless, starving or diseased. They also have the potential to destroy or damage stored comb. The greater wax moth can also cause damage to hive components by boring into the woodwork of the hives and frames and making boat-shaped indentations and is generally recognised as causing significantly more damage than the lesser wax moth.

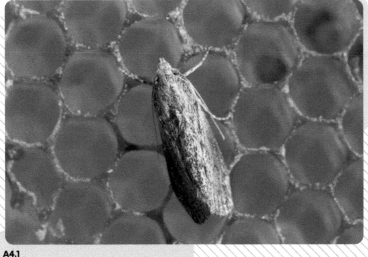

A4.1
Adult wax moth.

Do they have any benefits to bees and beekeepers?

In foulbrood diseased areas it is probably the 'beekeepers' best friend' as it removes infectious comb from the disease cycle. This is particularly important respecting feral or abandoned colonies. Certainly, when the greater wax moth arrived in New Zealand, as an exotic species, it was accompanied by a decline in the number of cases of American foulbrood.

What is the life cycle?

In order to control wax moth infestations it is important to be able to recognise them and understand their life cycles so that appropriate action may be taken.

Greater wax moth, Galleria mellonella

The adult moth has a length of about 20 mm, a wingspan between 24 and 33 mm. and is a brown colour with ash white markings. When seen in a hive it makes short runs or flights to darkness. It can sometimes be seen perching and flying in the vicinity of bee colonies at dusk, usually entering hives or boxes at that time. Females lay clumps of eggs in crevices within the hive, laying between 300 and 600 eggs which are pink/cream/white and are difficult to see. They hatch after 5–8 days into the larvae that cause the damage to bee combs. These larvae cannot ingest beeswax but eat it and live on the impurities contained therein. As a result, they are generally found in the brood comb or any comb containing organic matter. The larvae burrow through combs often just under the cappings leaving a silken white tunnel behind them. The bee pupae in the cells are rarely damaged, but sometimes become trapped in the cells by the silk threads and die. This condition is known as galleriasis and is more frequent in newly drawn comb. The larvae grow from 24–33 mm in length and when they pupate, often burrow into wooden frame components next to frame lugs, or adjacent to the hive walls leaving boat-shaped furrows about 15 mm long. In serious infestations the entire box can be filled with pupae in white silk cocoons. These are usually accompanied with dark specks of frass. On emergence adults mature and usually mate within the hive.

Lesser wax moth, Achroia grisella

The adult moth has a length of 10–13 mm, a wingspan of 11–14 mm and is silvery-grey to buff in colour with a yellow head. When seen it either flies, runs very quickly or holds onto the comb vibrating its wings. Each female can lay 250–300 eggs hatching into larvae that are similar in appearance to greater wax moth larvae but are smaller in size. Though larvae consume honey, pollen and wax they are not found in comb occupied by bees and do not damage hive components. Lesser wax moth larvae are unable to compete with greater wax moth larvae as the latter will eat them. Keeping strong and healthy colonies is the best prevention against wax moth infestations and if they are not controlled, infestations can rapidly multiply, being exacerbated in warmer conditions.

Cases of mistaken identity

The larvae of the small hive beetle (SHB), *Aethina tumida*, shows a striking resemblance to wax moth larvae and will consume beeswax, honey, pollen and brood, resulting in total devastation of bee colonies. The two key features used to distinguish SHB larvae from the greater wax moth larvae are the three pairs of legs near the head and the spines protruding from the dorsum. Further details can be found in the NBU leaflet 'The small hive beetle, a serious threat to European apiculture'.

Reproduced from: National Bee Unit APHA, National Agri-Food Innovation Campus, Sand Hutton, York. YO41 1LZ, UK.

A4.2
Wax moth damage.

Although I have never used it, treatment available for use in stored combs is B401 (Certan) is a biological wax moth control that gives 100% efficacy. B401 produces a natural toxin that attacks only the moth larvae. The larvae ingest the spores of B401 which then germinate in the gut and release the toxin which destroys the gut lining, killing the larvae. The reproductive cycle of the wax moth is therefore stopped.

Another treatment, again not one I have ever tried, is burning sulphur strips in a stack of supers and frames. Treatment is repeated every 4 weeks.

Storing comb especially in warmer climates increases the risk of wax moth attack. Storing in plastic bags is usually ineffective because wax moth eggs are often already present and storing outside only works if temperatures are low enough (many days of sub-zero temperatures).

> **NOTE**
> PDCB (paradichlorobenzene) crystals should definitely not be used because they are toxic to humans and honeybees and leave residues in wax and honey.

APPENDIX 5
BAILEY COMB CHANGE

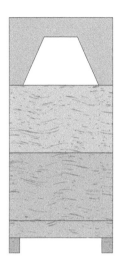

Week 1

1. Remove the queen excluder and any supers.
2. Remove any combs without brood and close up the gaps with dummy boards.
3. Place a clean box with clean frames filled with foundation, using the same number of frames as in the bottom box and ensure they are placed vertically above.
4. Put on a contact feeder with 2 : 1 sugar syrup.

↓

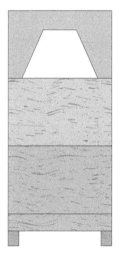

Week 2

1. Find the queen and put her in the top box, on a frame of eggs if possible.
2. Put a queen excluder between the two boxes. If possible, provide an entrance to the top box and close up the entrance at the floor level.
3. Continue feeding and add additional frames if required.

→

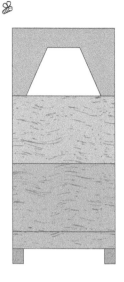

Week 3

1. Check for queen cells in the bottom box and remove.
2. If there are any frames with no brood in the bottom box remove them and close up the gaps.
3. Add additional frames to the top box to fill the box.
4. Continue feeding to draw out the combs in the top box.

↓

Week 5

1. Remove old brood box.
2. Remove and replace old floor.
3. Place new brood box onto the new floor.
4. Put a clean queen excluder on top of the brood box.
5. Add supers as required.

APPENDIX 6
SHOOK SWARM

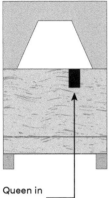

Queen in
a cage

1. Move the hive to one side.
2. Place a clean floor on the original site.
3. Place a clean queen excluder on the floor. (NB this is to stop the whole colony absconding.)
4. Put the new brood box complete with new frames and foundation on top of the queen excluder.
5. Try to find the queen and put her in a queen cage for safe keeping. If you can't find her, proceed to next stage.
6. Remove the middle five frames.
7. Taking each frame in turn from the old hive, shake all the bees into the gap in the middle of the new brood box.
8. Gently replace the combs in the middle of the box. Don't press them down!
9. Put the queen, in her cage, in between two of the frames.
10. Put on a crown board and feed with 2 : 1 syrup in a contact feeder.
11. After 48 hours release the queen from the cage if she is in one and remove the queen excluder from the floor.
12. Continue feeding for as long as is required to draw out all the new frames.
13. When all the combs are drawn out put on a queen excluder and supers.

If the shook swarm is for disease control, the old brood combs should be destroyed. If no disease is present and the shook swarm is being done just to transfer the bees onto fresh comb, then the brood frames can be put on the top of another hive to allow the brood to hatch.

APPENDIX 7
WAX PROCESSING

The system I devised to bleach wax using hydrogen peroxide was developed using trial and error as I could find nothing except vague references to it in any of the books I had. The chemical is easy to obtain on the internet, but at one time I learnt it was an ingredient for bomb making so not so easy to obtain on the high street.

A small quantity of peroxide is poured into a pan of molten beeswax. It is important that the wax is only just melted; too hot will result in the wax foaming over the top of the pan. The temperature can then be raised until the wax foams and is kept at that temperature until you think you have reached the right level of colour reduction. The problem now is to separate the wax from the peroxide mixture. Just pouring it out of the pan resulted in some of the liquid being poured with the wax. I solved this problem by taking the pan off the heat and standing it in a bowl of cold water. It should be left until it just starts to solidify, by which time the wax at the bottom of the pan will be solid. It can then be poured out leaving the peroxide underneath a layer of solidified wax.

The pan can then be put back on the heat and more wax added to be cleaned. It will foam again as the wax melts and more peroxide can be added as required. There is a slight residue of peroxide left, and this can be reduced by melting the wax on top of water and stirring it up and then letting it settle. The water will wash much of the peroxide out of the wax.

This *cleaned* wax is ideal for use in candle making, but there is sufficient residue of hydrogen peroxide that it is no use for making polish. It will corrode the tin! The aroma will also have been removed so it shouldn't be used for making foundation.

When using the wax for foundation it doesn't require purification to a great extent. Provided it is clean of debris it will be usable. I use two melting pans for this. One pan I melt the raw wax in and, having let it settle for a time after initial melting, I slowly pour most of it into the other pan which is used to keep the wax molten and to pour into the foundation press. Most of the dirt and foreign bodies will be left in the first pan which periodically I pour into a mould for reprocessing later. Surplus wax and trimmings from the foundation sheets can go straight back into pan number two.

APPENDIX 8
MAKING INCREASE

There are many books on queen rearing, and they all appear to make the process complicated. There is talk of cell starter colonies, cell raising colonies and cell finishing colonies. Various ways of transferring larva to queen cups, wax queen cups, artificial queen cups, grafting, double grafting. The list is endless and daunting to even the experienced beekeeper. In order to do the BBKA General Husbandry Assessment it is necessary to show evidence of queen rearing, so I persevered and after much trial and error developed a method that worked for me. I made many mistakes along the way, until I had developed a manageable method.

I didn't want to take any of my colonies out of production by making them queenless, even for a short period of time, but almost by accident I came across a method of raising queen cells in a colony, without removing the queen. The method had been detailed in the *American Bee Journal* in January 2002 and was developed by the National Bee Unit at Sand Hutton. The setup is shown in the diagram overleaf. It requires a big colony and is capable of raising 20 or more queen cells at one time. Grafted queen cups are put into the top of a double brood box, with a frame of pollen on one side and a frame of young brood on the other. Nurse bees, feeding the young brood will traverse through the frame with queen cells to reach the pollen on the other side. This theoretically means there is plenty of brood food being produced around the area of the queen cells, and the larvae are well fed. The sealed and hatching brood in the top box provide plenty of new nurse bees. The queen meanwhile is contained in the bottom box, away from the queen cells and is given some empty combs to lay in.

"I persevered and after much trial and error developed a method that worked for me. I made many mistakes along the way, until I had developed a manageable method."

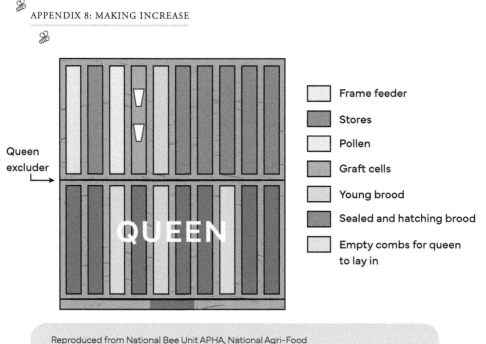

Queen
excluder

QUEEN

Frame feeder

Stores

Pollen

Graft cells

Young brood

Sealed and hatching brood

Empty combs for queen
to lay in

Reproduced from National Bee Unit APHA, National Agri-Food
Innovation Campus, Sand Hutton, York, YO41 1LZ.

I modified this system for use with my less prolific AMM strain of bees in conjunction with the Nicot Cupkit* system of larval transfer. It could be done using grafting and I used this method as well, as grafting means opening the hive fewer times, and it is not necessary to find the queen in the donor hive. Using the Cupkit method resulted in about 90% success, and with grafting it was probably about 70%. These figures might not be acceptable for the production of queens on a large scale, but it was good enough for me. The advantage with using the Cupkit is that the exact age of the larvae is known, whereas with grafting more guesswork is involved. I tried using the Jenter* system but without much success.

My system is to block off the top box so that there is room for just four frames, there is no feeder and the supers are put back onto the top. A queen excluder is put under the brood box as sometimes there is a tendency to

---

*   Cupkit and Jenter systems both use the same principle of containing the queen in a box with artificial cells. The queen is left in the cage, within the hive, for 24 hours. If she has laid eggs in the cage, which she normally will have done, she is released back into the hive. Three days later, once the eggs have hatched, the larvae, in their plastic queen cups, are transferred to the prepared cell bars and put into the cell raising colony. The exact age of the larvae is known and the disturbance to them is minimal. The colony does need to be opened up repeatedly over the course of a few days.

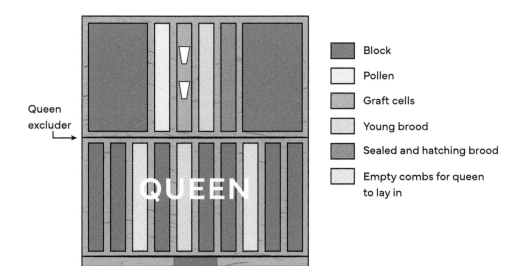

Queen excluder →

- Block
- Pollen
- Graft cells
- Young brood
- Sealed and hatching brood
- Empty combs for queen to lay in

QUEEN

swarm. Once the queen cells are sealed (day 9 after the egg is laid) each cell is then protected using a 'hair curler' cage. This prevents the bees building brace comb around the queen cells. The other major reason for using the cages is that it ensures that the queen when she emerges is contained. If your timing isn't quite right or you can't open the hive until they hatch, then you have lost nothing. You only ever let them go full term without cages on once!

A variation on this system that I tried was to use the blocks in the lower box, and thus restrict the queen to just four frames. This works equally well, but does involve finding the queen, whereas with only three frames plus the queen cells in the top all the bees can be shaken off the frames back into the bottom box before making up the top box.

This whole method can be completed without ever finding the queen if the larvae are transferred by grating once you master the trick and find which grating tool works best for you. Having chosen which queen you are going to breed from, open the hive, find a frame with very young larvae and graft the smallest ones that you can pick up into your artificial brown cell cups. Then just put the frame of grafts into the ready prepared colony. Most of my grafting was carried out using the top of a hive as a work bench!

A8.1
Blocked off top brood box with frame of artificial queen cells.

A8.2
Sealed queen cells showing 73% success.

A8.3
Queen cells with 'hair curler' cages.

Producing virgin queens is relatively easy using this method, and ten or so queens are normally more than enough for the small beekeeper. If you carry out an artificial swarm the queenless portion can usually be split into three or four nucs. All the queen cells that the colony has raised are removed and a new about to hatch, or just hatched queen inserted into each mating hive. The cell, or just hatched queen, is introduced in the cage she is in and then released by hand a few days later.

Alternatively, a nuc can be made up from a strong colony and a queen introduced. All you have to do then is wait and keep your fingers crossed that the queens get successfully mated.

Should you find you have a surplus of virgin queens, there are normally plenty of other local beekeepers who will take them off your hands.

There are plenty of ways of making increase without raising queens as described above. If you have carried out an artificial swarm, the 'old' portion of the hive will have plenty of queen cells. This queenless colony could easily be split into two or more nucs, each with a queen cell. Once the queens have hatched and started to lay, these nucs can be expanded into full sized colonies by the end of the season. Similarly, colonies that have swarmed can be divided up into nucs.

Producing mated queens in quantity is more difficult. The queens are easy to produce, but to get them mated each virgin queen has to be put in a small mating colony. I tried all sorts, from five or three frame nucs to mini hives such as Keiler or Apidea. All worked with variable success, but generally speaking the larger the hive the greater the success. Using five frame nucs is far more successful than using mini nucs, although those tiny hives do need fewer bees so don't take so many out of honey production.

APPENDIX 9
MORPHOMETRY

Morphometry is a method of establishing the strain of honeybees by examination of the veins of the wings. The method is to take a sample of 30 wings, scan them and analyse by computer. The results are then plotted on a graph. The analysis looks at two specific features described as the Cubital Index and Discoidal Shift. In the pictures below, Cubital Index is the ratio of 'a' to 'b', in *Apis mellifera mellifera* (AMM) this should be less than 2. Discoidal Shift in AMM is negative.

The pictures below show examples of the results from tests that I carried out in 2011, which allowed improved selection of breeder queens and the elimination of those colonies showing marked differences from native stock.

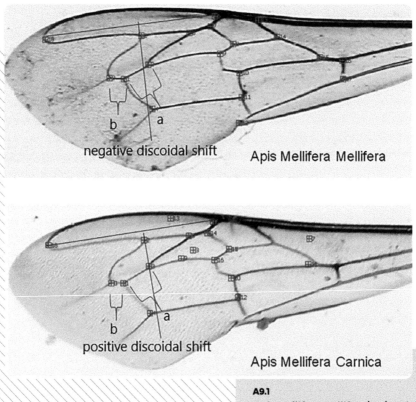

negative discoidal shift   Apis Mellifera Mellifera

positive discoidal shift

Apis Mellifera Carnica

A9.1
Apis mellifera mellifera (top) and *Apis mellifera carnica* (bottom) wings.

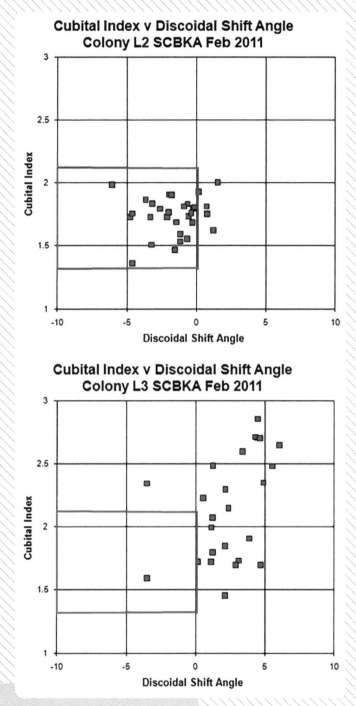

Cubital Index v Discoidal Shift Angle
Colony L2 SCBKA Feb 2011

Cubital Index v Discoidal Shift Angle
Colony L3 SCBKA Feb 2011

A9.2–A9.3
The two plots show an AMM result on the top
(with only a small amount of hybridisation) and
a non-AMM result on the bottom.

215

Index